NUMERACY AND MATHEMATICS ACROSS THE PRIMARY CURRICULUM

Building Confidence and Understanding

David Coles and Tim Copeland

David Fulton Publishers
London

This book is for Boo and Sally.

David Fulton Publishers Ltd
The Chiswick Centre, 414 Chiswick High Road, London W4 5TF

www.fultonpublishers.co.uk

First published in Great Britain by David Fulton Publishers 2002

The rights of David Coles and Tim Copeland to be identified as the authors of this work have been asserted by them in accordance with the Copyright, Designs and Patents Act 1988.

Copyright © David Coles and Tim Copeland 2002

British Library Cataloguing in Publication Data
A catalogue record for this book is available from the British Library.

ISBN 1 85346 640 9

All rights reserved. No part of this publication may be reproduced, stored in a retrieval system or transmitted, in any form, or by any means, electronic, mechanical, photocopying, recording or otherwise, without the prior permission of the publisher.

Typeset by Servis Filmsetting Ltd, Manchester
Printed and bound in Great Britain by Bell and Bain Ltd, Glasgow

Contents

Acknowledgements	iv
Introduction	1
1 Science: Quantifying the world	15
2 History: Mathematics and evidence from the past	25
3 Geography: A place for number	35
4 English: Patterns of language and patterns in language	43
5 Physical Education: Building geometrical thinking into playing games	56
6 Music: Notes and number	69
7 Design and Technology: Creating with numbers	80
8 Art: Keeping things in proportion	96
References	106
Index	107

Acknowledgements

This book, and our interest in mathematics across the curriculum, is a direct result of our early involvement with the Open University courses PME233 *Mathematics Across the Curriculum* and ME234 *Using Mathematical Thinking*. Our experiences with this course were fundamental in stimulating our commitment to, and understanding of, the vital role that mathematics can play in other subjects. Conversely, we began to appreciate how important it was for the mathematics to have a real-world context. It is apparent that many adults find difficulty in applying or identifying the mathematics they know in any subject not called mathematics. The material we present here attempts to identify the primary school mathematics that is, and can be, used in the other parts of the primary curriculum.

We need to thank, besides the original writers of PME233 and ME234, our colleagues at the University of Gloucestershire, especially Sally Palmer for help with the PE chapter, and Ros Swann, now at Nene College, for her advice with music. John Owens and Margaret Haigh of David Fulton have given us continual encouragement and advice, and have shown considerable forbearance.

Introduction: Why 'Mathematics Across the Primary Curriculum'?

Various justifications have been advanced for the place of mathematics in the primary curriculum:

- the importance of mathematics as an essential element of communication;
- the usefulness of mathematics in the everyday lives of adults;
- the uses made of mathematics in employment;
- the appreciation of beauty and satisfaction that can be gained through studying mathematics for its own sake;
- the value of mathematics as a means of promoting logical thinking and systematic working;
- the use of mathematics as a powerful tool for other subjects.

This book focuses on the ways in which mathematics contributes to the development of understanding in other subjects. It examines how the process of mathematical problem solving is consistent with, and contributes to, the development of problem-solving or enquiry skills in those other subject areas. Many of the other reasons for learning mathematics also contribute to our argument and understanding of the way mathematical thinking can be significant across the curriculum.

Problem solving and mathematical modelling

Although it might seem rather grandiose to label it as such, using and applying mathematics depends on the ability to identify an appropriate mathematical model. Abstraction of a mathematical model from the real world is not an easy task. Key features of a situation must be identified and relevant questions asked (see Figure 1). By stressing some features and ignoring others, an appropriate mathematical model can be constructed and answers calculated. For example, when five-year-olds are presented with a problem involving sharing 12 sweets equally between themselves and three friends, mathematical models must be constructed to answer the question, 'How many sweets will each child receive?' Most adults will first recognise the need to find the total number of children involved in the sharing and calculate $3 + 1 = 4$. Sharing will then be

1

matched to the mathematical operation of division and the calculation completed: 12 ÷ 4 = 3. Young children, however, will struggle to make these connections. This view of the mathematical modelling process is used in the Schools Curriculum and Assessment Authority (SCAA) paper on the teaching of number, which discusses the difficulties that pupils experience in using their understanding of arithmetic operations to solve numerical problems (SCAA 1997). The authors of the document suggest that a 'realistic' context can be beneficial in developing pupils' abilities to use and apply mathematical thinking:

> working with 'realistic' problems – problems that are 'sensible' in that they could have meaning outside the mathematics classroom – appears to provide sound learning experiences for pupils. Work at the Freudenthal Institute in the Netherlands has, over many years, demonstrated how working with 'realistic' problems can help pupils of all ages appreciate the need for mathematical models of situations and develop calculating skills.
> (SCAA 1997: 25)

THE MODELLING CYCLE

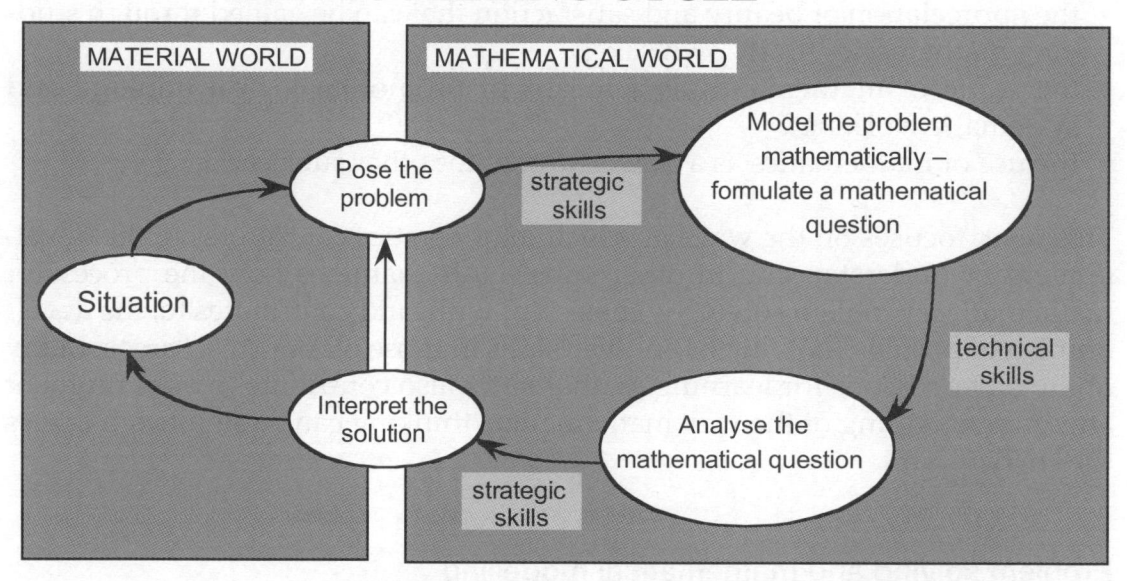

Figure 1 The mathematical modelling cycle (taken from Open University course ME234 (1989) *Using Mathematical Thinking*. Buckingham: Open University Press, Unit 13, p. 11)

Well-connected knowledge

In order to solve problems and use mathematics across the curriculum, pupils need to have a good understanding of the subject. But how can we be sure that a child has a solid grasp of the mathematics? For many, the extent of pupils' understanding can be seen in how they organise their own mental representations. Mathematical ideas, procedures or facts are understood only if they are part of an internal network in which they are linked to a variety of contexts and

Introduction

applied to solve problems. Greater understanding is achieved as these mental representations of mathematics are connected to form part of a network of representations. The degree of understanding is determined by the number and strength of these connections (Hiebert and Carpenter 1992: 67).

In the primary school this view of 'connected learning' has a bearing on how pupils are taught to experience and record mathematics. Examples might include:

- involving children in 'enactive' experiences in which the mathematics is tackled using objects (e.g. using bricks to represent a division calculation);
- using language that captures and describes the key mathematical elements (e.g. when you divide 12 children into 3 teams there are 4 in each team);
- drawing pictures and diagrams that record the significant mathematical ideas:

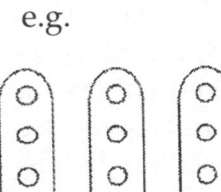

    e.g.

- using symbols that record the meaning more concisely (e.g. $12 \div 3 = 4$).

These ideas have a long history that can be traced from Dewey, Piaget and Bruner to Liebeck and Haylock. Experience, Language, Pictures, Symbols (ELPS) distinctions are used by many authors to think about different ways in which mathematics might be learnt (see Figure 2). Here we consider how these distinctions help us to think about the different ways of tackling the

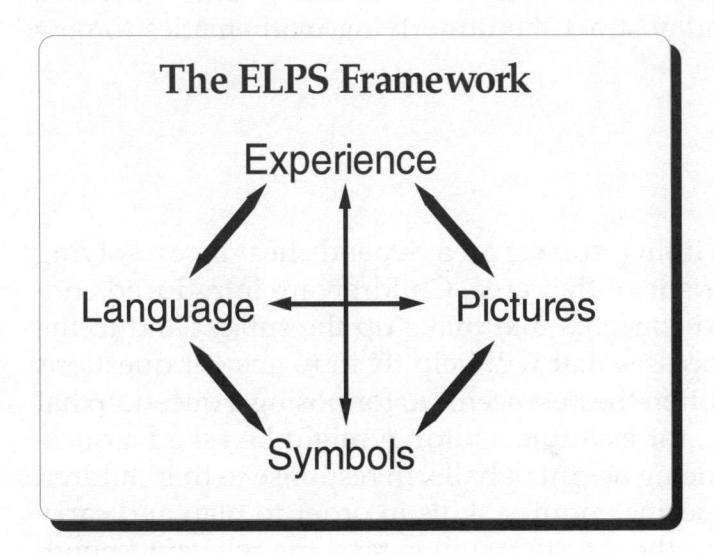

Figure 2 The ELPS Framework (Haylock and Cockburn 1997)

3

mathematical modelling process. Initially the child may be able to complete the sharing puzzle with real sweets and real children. Gradually, mainly through school activities, the child will learn to represent and interpret the problem with pictures and diagrams, and will gain the appropriate vocabulary to represent and communicate the concepts and to explain the skills involved in the solution; in time, the abstract symbolic way of representing the problem will also be understood.

Situated cognition and the transfer of learning

In the past it was usually assumed that the transfer of learning from one context to another was relatively easy to achieve. A general understanding of an abstract problem taught in a mathematics lesson could, it was thought, be used to solve problems in other contexts. Pupils would be able to extract the general mathematical principles from a few examples and learn to apply them in other contexts. In many situations, however, the reality was quite different. It soon became clear that pupils do not always see the relevance of the mathematical techniques they have been taught. Some may try to calculate an answer using an inefficient strategy, or may give an inappropriate answer based on the result of applying a mathematical technique uncritically. Current theories surrounding learning use the label *situated cognition* to refer to this effect, where the cognition or understanding is situated in, or limited to, a particular context. (For further discussion see Hughes *et al.* 2000.)

If transfer of learning from one situation to another is not as straightforward as was once assumed, learners will need to be shown multiple contexts and taught how to ask appropriate questions – including those that can be answered mathematically. It will also be necessary to teach the techniques needed to answer the mathematical questions that arise. Children need to encounter mathematical skills and concepts in a wide variety of contexts and the mathematical similarities in each one must be identified explicitly. Only then will pupils be able to generalise and abstract the underlying mathematics to new and different contexts.

Enquiry skills

The learning that takes place within a subject area depends heavily on solving the problems that are characteristic of that area. Children are introduced concurrently to the facts, skills and concepts that make up the subject and to the enquiry skills and thinking processes that will help them to answer questions that arise. A real-life context is often the best scenario for posing a question that leads to exploration. In science, for example, children might be asked a question about what affects the bouncing height of balls. In response to this children are taught to employ subject-specific enquiry skills in order to plan and carry out an investigation. Answering the question will extend the relevant knowledge and skills that the children already have and will help them to develop the

Introduction

technical skills to use particular apparatus or materials that are encountered in science.

For the ball bouncing investigation children will initially need to use their enquiry skills to plan a 'fair test'. They will then use their technical skills to explore methods of measuring, taking an average of several trials and putting the results into a table (see Figure 3). Having organised and analysed the information collected, children will generate answers to the original question that will again require subject-specific skills for interpretation. For our ball bouncing example these conclusions may involve statements about which type of ball bounced highest or which surface produced the lowest bounce for a particular ball. In many cases the answers attained lead on to further questions so that the enquiry cycle is re-entered. In science, this further questioning involves the development of theories that try to explain *why* the observed results were obtained.

THE PROBLEM-SOLVING CYCLE

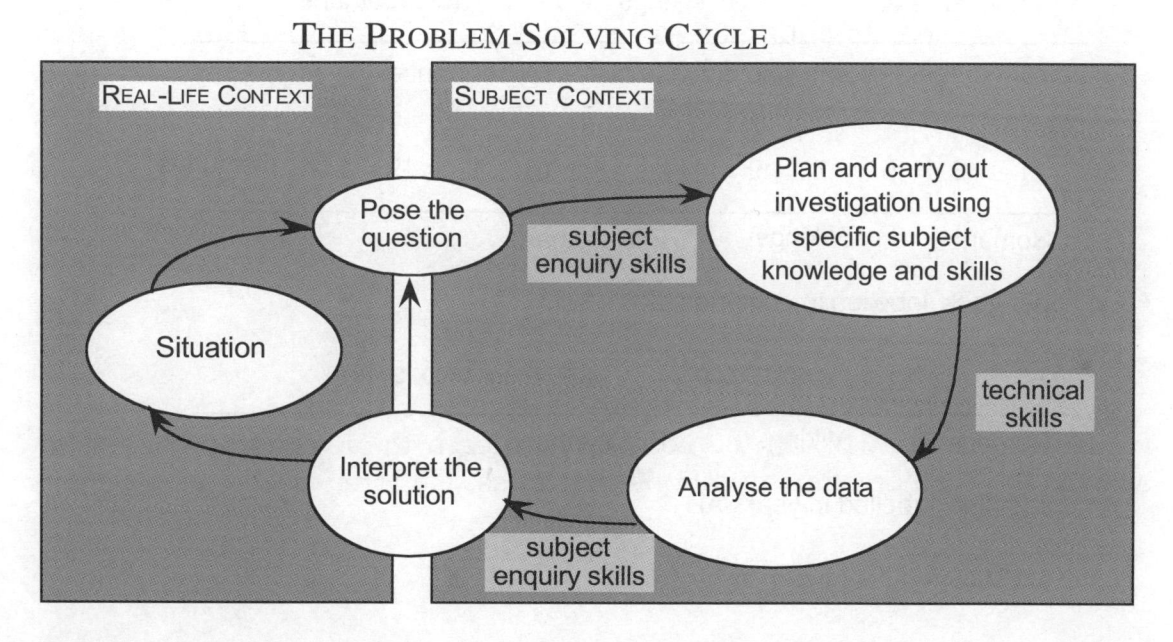

Figure 3 A general enquiry cycle

How mathematics enhances understanding in other subjects

Having examined a general model of the process of enquiry within a particular subject, we can now begin to identify the role that mathematics plays in developing understanding across the curriculum. Many subject areas rely on mathematics as a tool that underpins many of the key ideas of the subject. Even very basic mathematical concepts can be used to enhance understanding in the subject – a good grasp of the relative size of numbers, for example, is crucial to the understanding of chronology in history. Words like *century* and *decade*, and their relative sizes, need to be understood so that concepts relating to the passage of time can be grasped.

The sequence of historical events can be linked to an understanding of number order. Putting a set of events into their correct historical sequence is related to ideas about *more than, less than* and the *relative size of numbers*. For example, the chronological sequence for the periods of the Romans, Vikings, and Elizabethans could be memorised by connecting the periods to the sequence 1, 2, 3. It would be even more useful, however, to link this to a timeline using the standard measurement of years. This allows the possibility of understanding the relative lengths of periods of time, the concept of overlapping eras, the size of gaps between periods and the inclusion of further examples (see Figure 4). The distinction between AD and BC might be linked to ideas about positive and negative numbers. (A clear understanding of the way historians label years and the implications of there not being a year zero might lead to a different interpretation of when the new millennium started!)

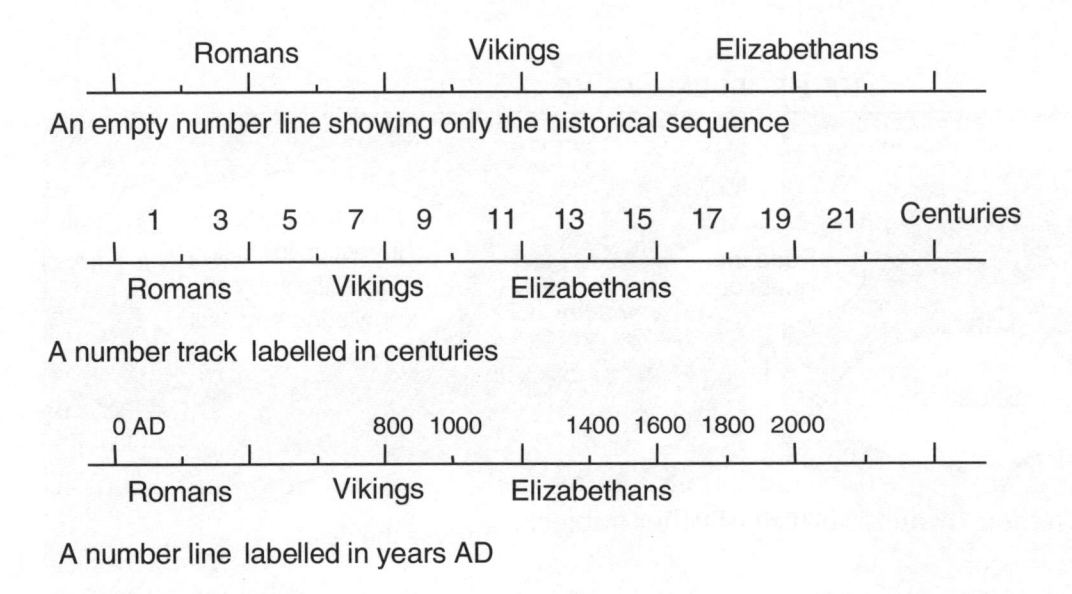

```
              Romans           Vikings           Elizabethans
   |_____|_____|_____|_____|_____|_____|_____|
An empty number line showing only the historical sequence


     1    3    5    7    9   11   13   15   17   19   21   Centuries
   |____|____|____|____|____|____|____|____|____|____|
    Romans      Vikings      Elizabethans
A number track labelled in centuries


   0 AD              800  1000      1400 1600 1800 2000
   |_____|_____|_____|_____|_____|_____|_____|
    Romans      Vikings      Elizabethans
A number line labelled in years AD
```

Figure 4 Using number lines in history

If children find the size of numbers like 1584 difficult to deal with mentally, counting in centuries rather than years may be a sensible simplification. (Note that the usual way of saying a year in history, 'fifteen eighty-four', will not help children to see this as a number.) This could be linked to concepts and strategies for approximation and estimating answers to arithmetic calculations. The addition of some prehistoric event to this sequence makes the context for the numbers involved much more challenging. Drawing timelines on a suitable scale to show how long ago dinosaurs lived gives a substantial impetus to developing ideas about the size of numbers, and can also contribute to pupils' ability to calculate and use measuring scales.

The important role that mathematics plays in helping children to understand other subjects is often overlooked or not explained clearly enough. This has implications both for the pupils' understanding and attitude towards mathematics and their grasp of other areas of the curriculum.

Introduction

The power to explain

One of the main reasons for learning mathematics is to help us understand the world and to identify and explain the patterns that we see in it. Many of these patterns can be understood more clearly by converting our knowledge of the world to numbers or by using simple sorting and classifying techniques. For example, in order to understand how patterns of rainfall vary between countries, it would be helpful to represent typical monthly rainfall levels on a bar graph. Once the data have been classified according to season, the general ideas about weather patterns and climate can be developed. Theories of situated cognition suggest that pupils who have encountered bar graphs in mathematics would not automatically recognise their application in answering geographical questions. Only with some prompting will many pupils see that they are, in fact, using mathematics and employing it in an appropriate way.

How other subjects enhance understanding of mathematics

A 'real-life' context, in which children can see how number relates to everyday objects, makes sense of the concepts behind the four arithmetic operations (addition, subtraction, multiplication and division). Undoubtedly, the use of apparatus specifically designed to help children understand aspects of number will also help with aspects of learning facts, developing skills and understanding concepts. However, it is through *using* mathematics in the context of other subjects that pupils develop their ability to apply mathematics. In these contexts they can develop their mathematical thinking strategies and learn to appreciate the role that mathematics plays in the real world and how it contributes to their understanding of other subjects.

Inevitably, much of the pupils' time in mathematics and numeracy lessons will be spent learning appropriate facts and skills. Nevertheless, teachers must be aware of the overall development required if their pupils are to become confident and competent users of mathematics. Concentration on facts and skills to the exclusion of the practical applications may well be part of the problem that some children have with mathematics, not part of the solution to the perceived mathematical incompetence of many young adults, as some would care to believe. When using mathematics to solve problems in other subjects pupils must be shown how to ask the mathematical question and interpret its answer in terms of the original question posed in the other subject. We think such teaching will have a significant impact on the pupils' understanding of mathematical concepts and skills and their ability to apply mathematical ideas to solve problems in the future.

How understanding is socially constructed through language

Ideas are constructed through interaction with others. Discussion helps children to master the language of mathematics and to clarify their conceptions of the subject. They can try out their ideas and solutions and develop their understanding while problem solving or using mathematics in other areas (e.g. approximation and accuracy when measuring the distance travelled by different vehicles made in design and technology (D&T)). Using mathematical ideas to explain aspects of other subjects will allow pupils to rehearse mathematical language in a new context. They can obtain feedback from other pupils or their teacher as to whether their explanations make appropriate use of mathematical vocabulary and concepts.

Cognitive structures and problem solving

Pupils' understanding is shaped by the problems they encounter. A full understanding will only arise when they face a wide range of problems. If pupils tackle mathematics in a restricted range of contexts the cognitive structures they build will also be limited.

The context of a problem may allow pupils an opportunity to judge the accuracy of their own answers. A pupil finding the difference between 3.84 m and 4.36 m may calculate an answer of 1.52 m in the classroom through the incorrect use of a standard algorithm. However, if they are finding out who jumped furthest and by how much during a physical education (PE) lesson, the clearer image of the length 1.52 m in this context may cause them to check their calculation and spot the error in their use of the standard algorithm. In this way the cognitive structures that the pupil builds in relation to subtracting decimals in mathematics will be extended by the use of subtraction and measuring in PE. This promotes better connected understanding that links contexts, experiences and language with mental images, which we are arguing is the basis of conceptual understanding.

Is there more to numeracy than mastering skills in arithmetic calculation?

Learning in mathematics occurs in a number of qualitatively different domains. The activities featuring in this book may challenge some of your beliefs about what mathematics is and how it should be taught. Some teachers view mathematics as a subject in which there are right and wrong answers. They regard good teaching as something that gives children the clearest explanations of the skills required to answer textbook questions correctly, so that pupils find it easy to reproduce these skills in the future. Askew has called this the *transmission model* (Askew 1999). Other teachers view mathematics as a creative subject in which pupils can be encouraged to find their own methods to solve problems; they regard good teachers as those who introduce children to appropriately challenging problems and support pupils as they develop ways to find solutions. According to Askew one of the key characteristics of more effective teachers of numeracy is the degree to which they encourage their pupils to see connections between different aspects of mathematics. We argue that in order to be really effective, the teacher needs to help pupils not only to make connections between ideas within mathematics but also to see the application of mathematics within other subjects.

Introduction

The current national guidance for primary schools in the UK makes some references to the importance of the application of mathematics but in our view does not give them enough emphasis. The National Numeracy Strategy approach to calculation emphasises the development of mental and oral calculation strategies, as well as written strategies. It also focuses attention on the development of understanding and the ability to solve realistic number problems. The National Numeracy Framework (DfEE 1999) makes some general comments about the usefulness of mathematics in other subjects but does not provide any explicit links.

While the National Curriculum (2000) for England (DfEE/QCA 1999) sets out the requirements on the use of the language and information and communication technology (ICT) across the curriculum, it does not include equivalent general instructions on the use of mathematics. The margin notes for specific subjects do indicate in a general way where links can be made to other subjects including mathematics. Under the heading 'Breadth of study', the Programme of study for mathematics gives a list of the contexts in which pupils should be taught mathematics; this identifies a variety of ways in which pupils might apply their mathematical knowledge, skills and understanding including 'using mathematics in their work in other subjects'.

Key skills and thinking skills

In discussing learning across the curriculum, National Curriculum 2000 does identify a number of key skills, one of which is *the application of number*. In explaining this further its authors refer to the development of mental calculation skills and the use of mathematical language as pupils 'process data, solve increasingly complex problems and explain the reasoning used'(DfEE/QCA 1999: 74). They go on to state that: 'Pupils need to be able to apply calculation skills and the understanding of number to problems in other National Curriculum subjects and to real-life situations' (p. 21). Another key skill is *problem solving* for which teachers are to provide opportunities in all subject areas. National Curriculum 2000 also identifies a range of generic *thinking skills* that are classified as follows:

   Information-processing skills
   Reasoning skills
   Enquiry skills
   Creative thinking skills
   Evaluation skills

The activities presented in this book meet the National Curriculum suggestions for developing mathematical skills in other subject areas and are designed to provide opportunities to develop the key skills of *number application* and *problem solving*. The book also explores the mathematical elements of the National Curriculum *thinking skills* and does so in a way that the authors have not found elsewhere, exploring in more detail the part that mathematical thinking plays in the development of these skills.

Types of mathematical learning

It may be helpful to see mathematical learning as something that is composed of a number of qualitatively different elements that interact as pupils develop their understanding. Cockcroft (DES 1982) and Her Majesty's Inspectorate (HMI 1985) used the following classification of learning objectives:

> Facts
> Skills
> Conceptual structures
> General strategies
> Personal qualities and attitudes

An alternative but related classification used in the SCAA (1997) discussion paper on the teaching of number at Key Stages 1–3 identified three general approaches to the teaching of mathematics:

1. Conceptual
2. Computation
3. Process

Using the former classification we argue that confidence and understanding amount to more than simply knowing mathematical facts and having the skills to perform some mathematical operations. Developing understanding requires an appreciation of the relationships between different concepts. These concepts must also be linked in and to a wide range of contexts. The ability to use and apply mathematics in context – in other subject areas in school and in solving real-life problems outside school – requires the development of mathematical thinking strategies and a positive attitude towards mathematics. In passing we welcome the fact that 'Using and applying mathematics' is not a separate component of the revised National Curriculum but is explicitly included in each profile component.

The latter classification (SCAA 1997) hints at the significance of teachers' underlying beliefs about the subject they are teaching. Our emphasis here is on the processes of mathematical thinking and the way these are relevant to and support problem-solving or enquiry skills across the whole primary curriculum. At the same time we are not ignoring the conceptual approach to mathematics teaching since we think the identification of mathematical concepts in other subjects will enhance pupils' understanding of mathematics. Furthermore, opportunities to practise mathematical skills in the context of other subjects will develop pupils' computational ability in a range of meaningful contexts.

How do pupils develop confidence and understanding in mathematics?

We shall now explore in more detail the ways in which stronger links between mathematics and other subjects can enhance pupils' understanding of both areas.

Introduction

The skill-using/skill-getting cycle
Teachers often have to choose between focusing on problems that use mathematical skills already mastered within the mathematics lesson or on problems that require the pupils to develop new skills. The relationship between learning skills and using skills is complex but can be seen to have an underlying cyclical nature (see Figure 5). Some might argue that when working in another subject area, children should be using skills that they have already encountered in mathematics lessons. The other subject provides a real-world context in which pupils can rehearse the skills they have already acquired, and the mathematics can enhance their understanding of the other subject. The two are in a sense complementary. One might say that in this case mathematical skills are *enabling* pupils to tackle problems that are at the same time giving meaning to the mathematics. Such an approach is sometimes characterised as 'learning the "basics" before applying them' and while it has a part to play in many situations, difficulties may arise from its narrow use. The SCAA discussion paper suggested that pupils find it difficult to make connections between the mathematics they have been taught previously and the problem context, and as a result may use inefficient methods to calculate an answer. It also identified that many pupils struggle to interpret the answers they give in the context of the problem; they focus on the mathematics and give a mathematically correct answer without realising that it is inappropriate in the context of the problem.

The skill-using/skill-getting cycle

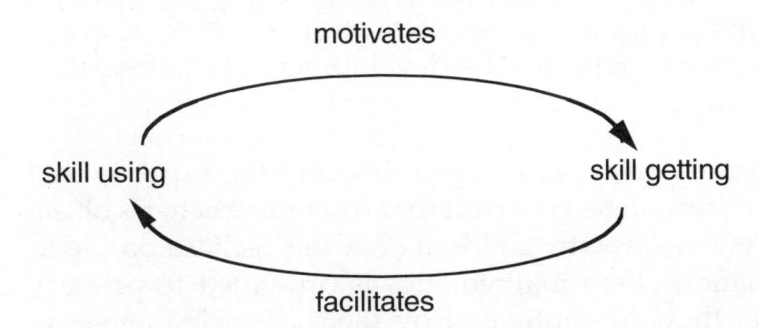

Figure 5 The skill-using/skill-getting cycle

Alternatively, if one uses work in another subject to focus pupils' attention on mastering new skills, the pupils will see that the subject has a purpose (a context) and will be motivated to learn new skills. By analysing the subject and identifying questions, pupils will inevitably come across problems that lead on to related mathematical questions, thus allowing mathematics to develop in the context of the other subject. The application of the mathematics is therefore an integral part of the learning process rather than an additional element in which 'problem solving' is learnt. The SCAA authors argued strongly in favour of this approach to using and applying mathematics:

> In the context of National Curriculum mathematics, this would suggest that learning activities that focus exclusively on skills are less likely to be successful than those which require pupils to integrate skills with knowledge and understanding in number and the 'using and applying' strategies of problem-solving, communicating and reasoning. (SCAA 1997: 26)

In either case we can argue that the work in other subjects is having a positive impact on pupils' attitudes towards mathematics and their understanding of its usefulness.

Attitude

Many adults have a negative attitude towards mathematics and leave school believing that they cannot 'do mathematics'. As a result they do not notice the mathematics that they frequently use in everyday life and at work. This negative assumption is widespread and held by many teachers and student teachers alike. The work of Laurie Buxton identified extensive dislike of mathematics among adults in the 1980s (Buxton 1981) and this was confirmed by research carried out for the Cockcroft Report (DES 1982). Among those beginning teacher training one finds many such negative attitudes towards mathematics (Crook and Briggs 1988). We think this negativity often reflects the fact that only token gestures are made towards applying mathematics to find solutions to real problems. In school mathematics lessons, the emphasis is mainly on mastering skills, with only occasional excursions into investigative exploration. While opportunities to identify mathematical elements of other curriculum subjects have often been overlooked, a more detailed examination of the potential advantages of doing so reveals a number of ways in which these benefits might enhance a pupils' attitude towards mathematics.

Having a purpose and motivation

Motivation is a very important component of learning and the purpose that pupils see in a classroom activity can be very different from the teacher's objective. There can be a variety of reasons why children or adults tackle a problem-solving activity in mathematics. Often mathematics is presented to primary children in such a way that they are motivated by their intrinsic interest in solving the problem. Sometimes the pupils' curiosity is stimulated by the problem itself, or by the skilled way in which the teacher introduces and initiates the activity with the children. Much mathematics can be covered in explorations within the subject itself and often a successful opening presentation of a task and pupils' natural curiosity are sufficient motivators for children and skilled teachers alike. In some cases, however, this interest is lacking: children complete the task because the teacher makes it obligatory and the task becomes 'work'.

When children are engaged in a task in a geography lesson and they become involved in asking real questions that mathematics can clearly help them to answer, they are more likely to struggle to understand and make sense of the mathematics. The problem in geography gives real purpose to using the mathematics. Haylock and Cockburn (1997) have suggested that the motivation of

Introduction

pupils is proportional to the degree of purposefulness that they perceive in the learning task. Although this sense of purpose might arise during their work within mathematics, it is also possible to harness the pupils' interest in other subject areas to ultimately enhance their motivation towards mathematics. Their curiosity may be activated initially by the geographical problem and then carried through into solving the mathematical puzzle. Although pupils may still need a degree of persuasion to complete the mathematical element of the problem, the curiosity derived from the subject area will give the process more purpose and will consequently motivate the pupils. The more involved they are with the question, the more motivated they are to master the mathematics and the more concerned they will be to check their thinking to ensure that their answers are correct. This opens up many more possibilities for pupil self-assessment.

Choosing mathematical tasks and initiating mathematical activity
The mathematical tasks chosen and the way they are presented have a great influence on how children's conceptions of mathematics develop and on their attitude towards the subject. These choices and decisions reflect the teacher's own understanding and beliefs about mathematics. Where the teacher sees and presents mathematics only as a separate subject, pupils will tend to develop less well-connected understanding of the mathematical skills and concepts that they encounter and will therefore be less able to apply the mathematical skills they do have to solve problems. Conversely, teachers who make explicit reference to mathematics as it is used in other subject areas help children to build better connected mathematical knowledge and to see its relevance to solving problems and understanding the world.

Why identify mathematics in or through other subjects?

In this book we explore how teachers of individual subjects can make effective use of mathematics to enhance pupils' understanding of both the subjects and the mathematics. Many people have recommended contextualising mathematics in order to help children make sense of it, but this is not an easy task and may actually have the reverse of the desired effect. The intention of this book is to identify contexts in which mathematics is used, to explain and advance ideas in other subject areas so that pupils' understanding across the curriculum is enhanced. Many examples of applications of mathematics in published mathematics schemes are selected to explain the mathematics, often in ways that over-simplify the subject and make the mathematics seem peculiar and irrelevant. Multiplication worksheets that use drawings of animals to put ideas into a context will not set problems for which pupils see a real purpose. Tackling the sections of work on money and decimals by setting questions on choosing and buying presents may have no useful function for nine-year-old pupils who are told by their parents what presents to buy for their friends. This approach may well lead to pupils having a poor attitude towards mathematics and is likely to encourage many to ask the question, 'When is this mathematics useful?' The hidden curriculum message of many of these examples seems to be that mathematics is about answering questions that real people don't ask and which bear no relevance to real life.

This book aims to provide the teacher with a sample of activities in different subject areas that illustrate key applications of mathematics in context. We have argued that the use of mathematics in other subjects – even at the most basic level – will provide a much firmer basis for pupils to develop problem-solving skills and a connected knowledge of mathematics – both of which are at the core of confidence and understanding. For pupils encountering mathematics in other school subjects, the transfer of learning between lessons must be simplified and the weaknesses of single context learning reduced. From the examples given it should be possible to develop more of your own activities that fit your school context and pupils more closely. We hope these activities will provide exciting and intellectually stimulating ways in which to involve your pupils in using and learning mathematics.

This introduction has examined the thinking that lies behind the use of mathematics in the context of other subjects. We have explained how important it is to do more than just teach facts and skills if children are to become adults who are confident users of mathematics. The following chapters are divided into individual subjects and their structure is based on the problem-solving cycle shown in Figure 3. Each chapter opens with an examination of the subject context and a consideration of the aspects of the real world that the subject encompasses and the type of enquiry and problem solving it invites. This is necessarily brief and straightforward but is needed to set the scene for the discussion of the mathematics being used in subjects at primary school level. We have generally followed the structure of the National Curriculum 2000, which serves to identify learning objectives in terms of specific subject knowledge and skills for pupils, but the ideas should be more widely applicable. The role played by mathematics in the subject is explored in relation to the main concepts and skills of the subject area, and a range of activities are introduced that typify the use of mathematics within the subject's component concepts and skills. The activities are organised around the following headings:

- *Key concepts* of the problem-solving activity and how they relate to subject structure;
- *The mathematics* involved and how it is relevant to the activity and its objectives;
- Step-by-step description of the activity;
- *Extension activities* designed to reaffirm skills and promote further learning.

Each activity uses the subject context to develop problem-solving thinking within both the subject and mathematics. We think that every activity will involve pupils in using and applying their mathematics in some aspect of problem solving, communicating or reasoning mathematically. Identifying these instances in *every* activity would be unduly repetitious and we have therefore only explicitly identified the ways in which specific skills and concepts from number, shape, space and measures, and data handling are being applied.

1 Science
Quantifying the world

A key guiding principle for science is that rational enquiry can enable us to describe and explain the physical world and that this understanding helps us to predict and control aspects of that world. Consequently the study of science for primary pupils focuses on the description of physical processes and explanations of events in the physical world, through to the exploration of the key technology that we use in our everyday lives. The subject aims to bring system and order to our understanding of the world through the discovery of general laws that explain the way that things happen. The field of knowledge covered by National Curriculum science includes the study of the Earth and its position in the universe, living things, physics and chemistry.

Learning about the scientific method – the process of being scientific – is usually thought to be as important as teaching specific content in science. Today, the most widely accepted process of investigating scientifically currently starts with the generation of a hypothesis. It then moves on to designing an experiment to *disprove* the hypothesis and concludes with the establishment of some provisional knowledge about the world. This process requires the ability to generate a variety of hypotheses during an initial brainstorming phase. At this stage, teachers can support and stimulate children by asking questions based on the knowledge and experiences that they know their pupils already have or by asking them to observe closely some particular demonstration. Some of the ideas produced may be discarded immediately but a pupil's tentative explanation can often grow into a hypothesis that is worthy of development. If the results of a fair test do not contradict the initial predictions then it is deemed 'correct' until some future event or experiment leads to doubts and further revisions.

One of the first questions to arise early on in the process is 'What is happening here?' In order to find the answer, children must develop close observation skills and, on the basis of the information gathered, identify relationships and generate ideas that will enable them to answer the question 'Why is that happening?'

Knowledge base of the subject

The National Curriculum for science (DfEE/QCA 1999) divides the curriculum into learning about the processes of investigating in science through planning, collecting evidence and evaluating the results. It goes on to identify specific scientific content, covering knowledge of facts, learning skills and understanding concepts related to life, materials and physical processes. Under the heading of 'Scientific enquiry' a number of connections to the mathematics curriculum are revealed. These relate to understanding and using measures and to processing, representing and interpreting data. The 'Life processes and living things' section covers animals (including humans) and plant life, as well as ideas about classification systems and some aspects of ecosystems. The 'Materials and their properties' section explores the grouping and classification of materials on the basis of their properties and deals with key ideas about how materials change. Under the heading of 'Physical processes' pupils are introduced to the scientific study and understanding of electricity, light, sound, forces and also to Earth's place in the solar system.

Much of the knowledge gained by pupils will be evident in the vocabulary they use to describe the events they observe and to explain why those events occurred. Their ability to predict what will happen under new conditions is a good way of assessing the extent to which they understand scientific concepts, but knowledge will also be evident in the language chosen to express their ideas.

Where is the mathematics in science?

Mathematics is essential to the development of scientific understanding and all aspects of mathematics may be encountered in some aspect of science. Furthermore, the process of conducting a scientific investigation and the process of developing a mathematical model differ very little. As the investigation process in each subject is so similar, a scientific experiment can provide a good opportunity to identify the underlying mathematical reasoning and to discuss the process of problem solving. At the same time it provides a context for the application of specific mathematical skills and concepts.

The requirement to observe phenomena closely in science will give opportunities to develop understanding of classification systems, measurement and number. The need for increasing accuracy in measurement will be the key to pupils realising the importance of developing their understanding of number to include fractions and decimals. In order to discover patterns in their answers, pupils will need to interpret information carefully and order results in tables, identify relationships in words and possibly with formulae. In identifying relationships, pupils will use graphs in their analysis and to communicate their understanding. The mathematics of shape and space is particularly significant in work on materials and their properties, but will also be relevant to representations in many other areas.

Science: Quantifying the world

Activity 1

What materials are good for making a winter coat?

Key concepts
Classification plays an important part in solving scientific problems. In many topics at both Key Stages 1 and 2 sorting and classifying provide the key to developing understanding. Knowing the properties of materials will be a major step in solving problems and this knowledge depends crucially on the process of classification.

The mathematics
Understanding classification systems and recognising whether systems involve inclusive or exclusive categories is difficult but important. The use of Venn diagrams (or Carroll diagrams) gives a clear pictorial representation of how materials are classified using particular categories. For this problem a Venn diagram will provide a way of identifying which materials have all of the properties thought necessary for a winter coat. More sophisticated problem solving involves the use of rating scales and tables as a way of measuring the suitability of a material for a particular purpose.

Step 1
The children brainstorm to identify a range of properties that the material needs to have to make a useful winter coat.

Step 2
Reduce the number of properties identified to a suitable amount by focusing on the *key* features of winter coats. In KS1 this may be as little as two, but older pupils can deal with four or five at a time.

Step 3
Present the children with a range of potential materials and devise a way of testing them for the properties identified. At the simplest level the material should be judged according to whether or not it has the properties required to make a winter coat. More sophisticated testing may give each material a rating or a measurement score for each property.

Step 4
The children carry out the testing and record their results in a table.

Numeracy and Mathematics Across the Primary Curriculum

Material	Waterproof	Strength	Warmth	Flexibility
Plastic				
Cotton fabric				
Woollen fabric				
Leather				

Table of results

Step 5
When all testing is complete, transfer the results to a Venn diagram locating each material in the appropriate section of the diagram. A Carroll diagram may be used if there are only two or three properties under consideration.

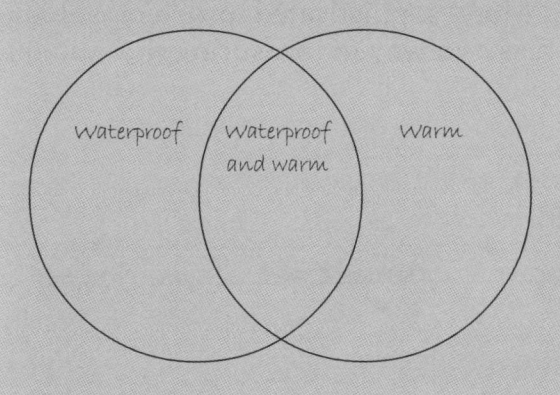

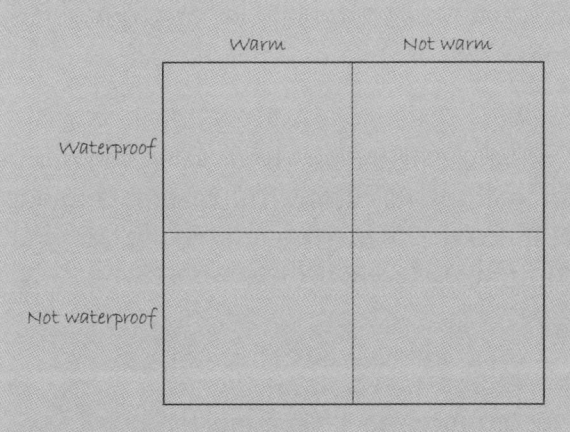

Venn diagram (Waterproof | Waterproof and warm | Warm)

Carroll diagram (Warm / Not warm × Waterproof / Not waterproof)

Step 6
Help the children to interpret their diagrams to decide which materials are suitable for making winter coats.

18

Extension activities
- Include more properties of materials in the testing.
- Use a rating scale to gauge the performance of each material in relation to each property. This makes the selection of the *best* material more difficult. The use of weightings can highlight the importance of some properties over others.

Activity 2

The Earth and beyond

Key concepts
Interpreting models, pictures and diagrams of the solar system will open up a whole new world for pupils and will give them a greater understanding of the relative location of the planets. Encourage children to appreciate the vastness of space by drawing diagrams to represent distances between the Earth and other parts of the solar system (e.g. draw a diagram to show how far the Earth is from the moon). See if they can work out how long it would take a spaceship travelling at 100 mph (or kph) to reach the moon.

The mathematics
This problem will involve division by 100 and will give children a concrete example of how division can be used to solve a real-world problem. The relationship between distance, speed and time needs to be understood in order to identify the calculation needed to find the answer. The appreciation of large numbers will be developed and it will be possible to introduce the use of rounding. For speeds of 100 mph it would be inappropriate to use a calculator, but if the activity is extended to consider other speeds then calculators may be introduced. They can also be used to complete the conversion of hours to days, weeks or months.

Step 1
Identify that the moon is a long way away and start some discussion about how far.

Step 2
Get the pupils to find the actual distance or alternatively tell them (240,000 miles or 384,000 km).

Step 3
To help children appreciate how far this is, generate a table of journey times to a variety of places travelling at 100 mph (e.g. London, Paris, New York or Sydney). Give the times in hours first but also convert them to days, weeks and months.

Numeracy and Mathematics Across the Primary Curriculum

> *Step 4*
> To show the relative distance, construct a diagram where one unit of measure equals the time taken to reach New York or Sydney and record how many times longer it takes to reach the Moon.

Extension activities
- Explore distances between the Earth and other planets in the solar system, or even other stars.
- Ask children to imagine they are travelling at different speeds: 50 mph, 500 mph or 1,000 mph. Use the concept of ratio to calculate the time taken from the results for 100 mph.
- As well as modelling the relative distances make scale representations of the planets according to their diameters or their mass, so that the sizes of different planets can be compared.

Activity 3

My Mother says I need to eat more fruit. Is she correct?

Key concepts
Calculating the types of nutrition in different meals and making a plan for a balanced diet over a week will allow children to consider how food is needed for activity and growth, and the characteristics of a varied and balanced diet. Start from the premise that views on what constitutes a healthy diet will change as the children find out more about good diet and they will need to be able to replan their own diets accordingly.

The mathematics
In this activity a variety of calculations will be needed to solve a number of real-world problems. The complexity of the calculations can be controlled by the way in which the information is presented to the children. For example, where information on the fat content of food is presented as the amount of fat per 100 g, the calculation will be easier than if the same information was given as a fraction or percentage, where the calculation would require an extra step. Pupils should cover multiplication and division calculations involving decimals and encounter the different uses of fractions, percentages and ratio.

Handling data can involve the presentation of information obtained in a variety of formats. Pie charts can be used to show the proportion of the major food groups in a particular meal, but the comparison of different meals or foods is best made using bar charts to show the proportion of a food group in different foods/meals. A bar chart of the recommended daily intake should also be drawn to help children make comparisons. Presenting data in this way will teach pupils how best to draw conclusions from the statistics and the graphs.

Science: Quantifying the world

Note that all these graphs must show the proportion of food groups in any one meal – expressed as a fraction or percentage – or weights if they are all adjusted to be for 100 g of food.

> *Step 1*
> Collect information from labels on food packets or advertising material, and produce bar graphs of the major food groups in different meals: record them all using a similar scale by calculating the quantity per 100 g from the information given. The proportion of a food group in a particular meal could also be presented using fractions and percentages.
>
> *Step 2*
> Match this to graphs of the recommended proportions of different food groups in a balanced diet. Which is the best individual food?
>
> *Step 3*
> Calculate the quantity of the major food groups obtained from a complete meal. This involves knowing the weight of the meal or the individual foods in the meal and the amount of protein per 100 g. Different groups of children can work on different meals.
>
> *Step 4*
> Present the information in tables and graphs.
>
> *Step 5*
> Compare these graphs to the recommended proportions of different food groups in a balanced diet. Which is the best meal?

Extension activities
- The values on packaging are means – how much do they vary?
- Ask children to record information about their eating habits over a week and use the data collected on individual meals to calculate their intake of the major food groups over the seven days. The pupils can then represent this information in a bar chart showing absolute quantities in grams and as fractions or percentages. Which way makes it easiest to compare their diet with the recommended balanced diet?

Activity 4

Toy vehicles: What makes the car go further?

Key concepts
In the real world, the higher up a slope a vehicle starts the greater the energy that the force of gravity is able to give to the vehicle. Pupils must learn about

the concepts of gravity and force as a way of explaining what happens to make the car move down the slope. They need to identify the relevant factors and devise (or be given) experiments to help them identify these.

The mathematics
This activity involves measuring in non-standard and standard units, choosing a suitable measuring instrument and recognising the approximate nature of measurement. The pupils can use the concepts of 'average' and 'range' to compare their results for different sets of data. If enough trials are carried out the pupils can use the mode to represent a series of results for a particular starting height. Alternatively they could calculate the mean using a calculator, or comparisons can be made on the basis of the range of results obtained for each starting height. The pupils should recognise that there is considerable variation in the result for each trial and that using only one trial at each starting height would be unreliable. The data can be collected in tables and then presented in line graphs. The final stage of the activity will involve the interpretation of graphs and tables to answer the original question.

Step 1
Provide children with a ramp and a selection of toy vehicles.

Step 2
Identify the factors that will be investigated which might make the vehicle go further.

Step 3
For a straightforward investigation of gravity you may restrict the factors tested to the distance up the ramp at which the vehicle is released.

Step 4
Devise a way of changing and recording the angle of the ramp.

Step 5
Carry out a number of trials and use a table to record the distance the vehicle travels for different slopes. The distance needs to be measured consistently (e.g. from the bottom of the ramp to the back wheels of the car) and the vehicle needs to travel in a reasonably straight line. The release of the car at the top of the ramp also needs to be consistent. If pupils already have good investigative skills they can identify for themselves the conditions required for a fair test. The measurement of the distance the vehicle travels can be in standard or non-standard units.

Extension activities
- It would be valuable to compare the effect on the vehicle when the slope is made from different surfaces. Introduce the complication of the force of friction.

- Ask pupils to try pulling a margarine tub with weights in it up a slope. How much force does it take to move it? Get them to measure the force required by using weights to pull the tub up the slope or with a force meter. How much extra force is required for every 10 g added to the weight in the tub? They can then draw bar graphs to identify the relationship. As well as testing different surfaces for the slope, children can also use different materials for 'tub', vary the angle of the ramp, or make alterations to the toy vehicles.
- What factors affect how high a ball bounces? This may be a different context but it requires the same measuring and data handling skills.

Activity 4

How can we keep the water hot?

Key concepts
In this activity children are being introduced to the idea of temperature as a measure of how much heat an object has. They will use measurements of temperature to assess the effectiveness of their methods of keeping the water hot. They may also encounter the changes that can occur in materials as they are heated and cooled. The activity can be used to develop pupils' ideas about properties of materials that act as good thermal insulators. The use of such materials has a wide range of applications including the retention of heat and cold.

The mathematics
Here the pupils' work on measuring temperature will give them an opportunity to practise reading scales with increasing accuracy. They will need to measure time accurately as they monitor the change in temperature over a given period. The data will initially be collected in a table before being turned into a graph and children will be given the opportunity to consider whether the data are best represented as a bar chart or a line graph. Where both of the variables measured are numbers it is better to draw a line graph. Comparing the graphs for different methods of keeping the cans warm will give the pupils a chance to interpret the graphs and draw conclusions.

Step 1
Pupils may need to practise reading the scale on the thermometer and the timer used, whether this is a clock or a stopwatch. There will also be safety issues concerning the use of the thermometer and hot water.

Step 2
Explain the problem by suggesting a situation in which hot drinks or food need to be kept hot for a period, for example at a school outing, a school fair or sports day.

Numeracy and Mathematics Across the Primary Curriculum

> *Step 3*
> Identify a range of materials that might be used to make a container for keeping a can of water hot.
>
> *Step 4*
> Each group of pupils design their own insulation system. Limit the number of different materials that may be used to either one or two, or perhaps fix the constraint by setting a maximum size for the container. The design for the container must include provision for inserting the thermometer.
>
> *Step 5*
> Each group make their container.
>
> *Step 6*
> Test each container by starting with the same quantity of water at the same temperature (40°C). Make all necessary safety checks at this point. Take temperature readings every minute and record in a table.
>
> *Step 7*
> Continue recording temperatures for at least 20 minutes or longer if possible.
>
> *Step 8*
> Draw graphs from the tables of data. Compare the graphs for different containers and decide which ones perform best at keeping the water hot. You might also include data from a can that is not insulated at all.

Extension activities
- Use control technology equipment to monitor the temperature (e.g. a hand held data logger or sensor attached to a control box linked to a computer). This simplifies the accurate monitoring of temperature change over a longer period of time.
- For a simpler experiment involving similar mathematics record the time that an ice cube takes to melt in water that is at different starting temperatures.
- You might identify what temperature the water will eventually reach (room temperature) and try to use the line graphs to predict how long this will take. Ask questions like, 'Will the water still be hot in the morning?' 'After lunch?'

2 History
Mathematics and evidence from the past

History is the construction of 'pasts' from evidence that has survived over a period of time. It is a systematic attempt to describe and explain the past, but is distinct from it, as only a selection of the events can be described and analysed due to the problems relating to the differential survival of evidence and the need to make some selection from the huge database of facts. By using the range of surviving evidence and asking questions of it, it is possible to put events in order, suggest their causes and get a sense of the period.

Where is the mathematics in history?

Usually the emphasis on mathematics in history is limited to timelines and chronology. However, mathematics has a clear role to play in enabling children to quantify primary evidence, and to help them draw conclusions and communicate their findings (Copeland 1992). While the setting up of appropriate historical questions, the choice of evidence and the conclusions drawn are all activities related to historical thinking, mathematics can aid this process by providing techniques and ways of thinking that support and refine historical thought.

Activity 1

Relative chronology: In which order did these events happen?

Key concepts
Chronology is the ordering of historical events and changes, based on the relative sequence between them. An understanding of number and simple arithmetic calculations, as well as many aspects of measuring, underpin ideas about time.

The mathematics
Relative chronologies are concerned with putting events in order without recourse to dates, which in some cases may not be known, and when only the sequence of events is significant, for example in prehistoric times before writing

was invented. Relative chronologies are also valuable for using with documents that describe the unfolding of an event, where no timescale is given but the actions in a sequence are connected to each other in time, for example the Roman Invasion and Boudicca's Revolt. Ordered lists, networks or flow charts can be used to represent these relative chronologies and enable the sequence of events to be put in the right order. Essentially, events are recorded in the form of 'this happened, then this happened'. The top of the flow chart will always record the earliest event, the bottom the latest.

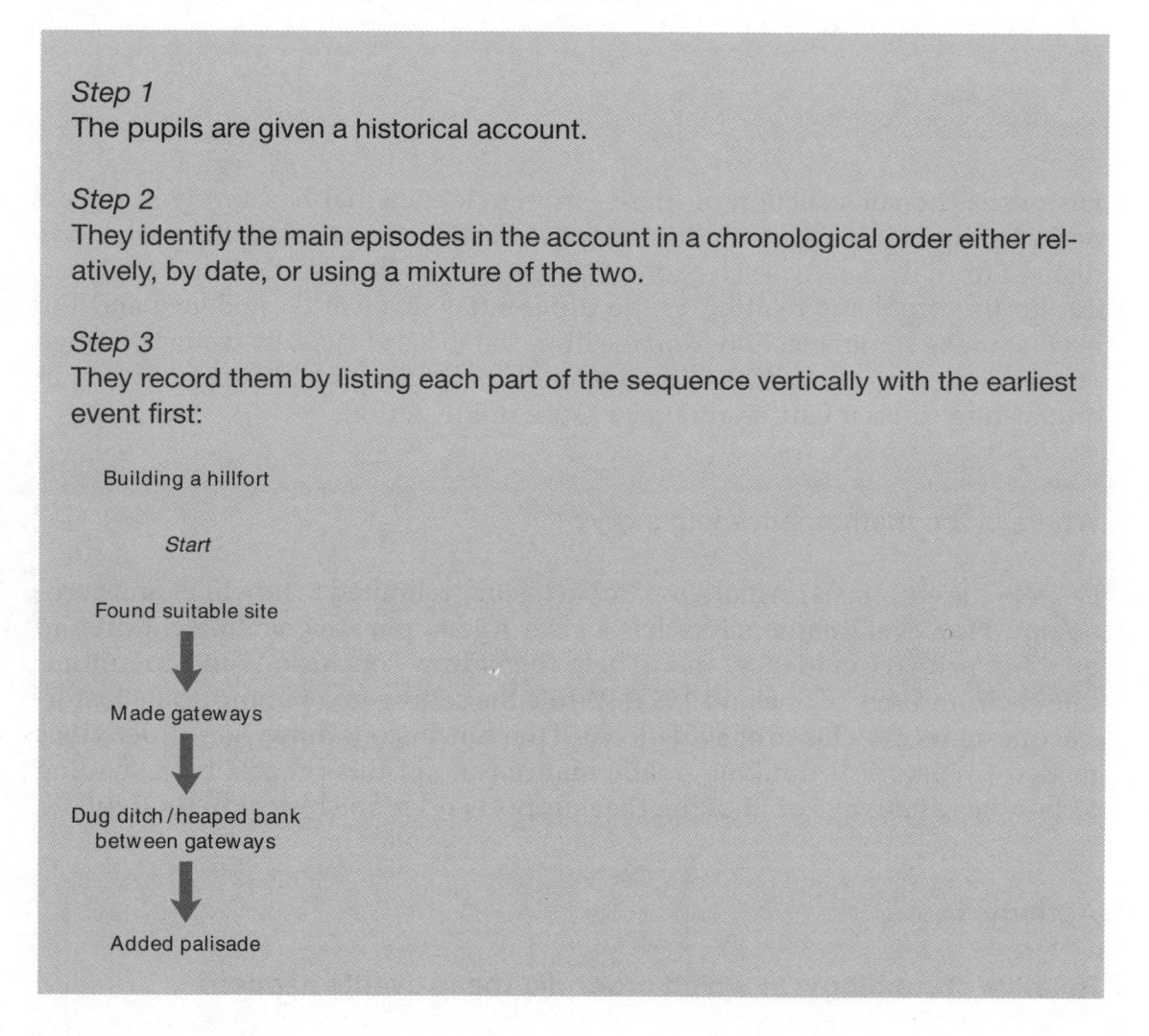

Step 1
The pupils are given a historical account.

Step 2
They identify the main episodes in the account in a chronological order either relatively, by date, or using a mixture of the two.

Step 3
They record them by listing each part of the sequence vertically with the earliest event first:

Building a hillfort

Start

Found suitable site
↓
Made gateways
↓
Dug ditch / heaped bank between gateways
↓
Added palisade

Extension activities
- Help pupils to make parallel flow charts with two events or buildings that overlap in time and are connected.
- Challenge them to make flow charts of the actions of two or more people in a historical event. The earliest action might be a shared one: 'the army arrived at the river...then someone did this... and another person did that . . .' The result will be two parallel flow charts starting from the same event, perhaps converging at various points, and joining again in the final event.

History: Mathematics and evidence from the past

Activity 2

Absolute chronology: What changed in my life and when?

Key concepts
For absolute chronologies where dates are known, or important to record, timelines can be used. Timelines are devices that use nominal scales for recording events and they focus attention particularly on change. As there are a number of different scales used for measuring and recording time (days, weeks, years, decades, centuries, geological time), the skills of choosing the appropriate scale and recognising the relative sizes need clear teaching and practice. When making and interpreting timelines, children will use skills in ordering large numbers, in mental and written calculation to obtain both exact and approximate answers, and in counting, multiplication and division by 10, for example using multiples of 10 to construct scales.

The mathematics
As well as ordering numbers there will be opportunities to use mental calculation strategies – for addition and subtraction – to calculate the periods of time over which changes took place. If dates are placed accurately on the timeline then there may also be an opportunity to use fractions (tenths) to locate numbers like 1066 on the timeline.

Step 1
Children make a list of the important dates in their lives.

Step 2
They record the changes that occurred in relation to these dates.

Step 3
They select the most appropriate scale for the timeline.

Step 4
They map their data onto the timeline.

Step 5
They record the changes.

Extension activities
- Construct the timeline on other scales covering longer periods of time, using a scale of 1,000 years to 1 metre. Help the children calculate how long the timeline would have to stretch back to the age of the dinosaurs (200 million BC) or some geological period in the distant past. On the scale suggested above (1,000 years to 1 metre) 1 million years will require 1,000 metres or 1

kilometre! Challenge the children to identify a scale for a timeline on which this length of time could be shown. Identify what length represents the last 2,000 years from zero to AD 2000.
- Use timelines to show the coincidence of events or periods, for example with British history on one side of the line and the comparative world history on the other. This is particularly useful in the case of pre-Columbian civilisations in the Americas, which are frequently referred to as 'Ancient' and consequently equated, in terms of chronology, with the Ancient Egyptians or Greeks, when in fact cultures such as the Aztecs and Incas are contemporary with the early Tudors in Britain.
- Work with pupils to record changes in the local community on timelines about the school or locality. Compare local history with events and changes at a national level.
- Use complex timelines to record different chronologies: Christianity, Islam, Judaism, etc.

Activity 3

How do we count BC?

Key concepts
Children are often confused when dates BC are used to refer to events from the period before the year once thought to be Christ's birth. They find it hard to grasp the ordering of such events. This activity focuses on using a timeline to help pupils understand the sequence of events and the number of years between episodes.

The mathematics
The key mathematics of this activity involves extending the number line to include negative integers, being able to order a set of these and calculate the difference between two integers. The link between history and work with negative integers on a number line is complicated by the fact that historians do not use a Year 0, so calculations of differences across years AD to BC give results that are one less than the same calculation with integers!

Step 1
Pupils are given a list of events without dates and are asked to find the dates using resource material.

Step 2
They use a prepared number line for dates BC and mark the events on it.

Step 3
They make an ordered list of the events from earliest to latest.

History: Mathematics and evidence from the past

Step 4
In discussion with the teacher they notice how this differs from dates AD where larger numbers mean the event was later. For dates BC the larger number means the event was earlier.

Step 5
Pupils rehearse using the language correctly by asking and responding to questions about the sequence of events. All pupils need to be given opportunities to say sentences like, 'X happened in 340 BC and took place before Y, which happened in 240 BC'.

Step 6
Pupils ask each other questions about the length of time between events and use their timelines to help calculate answers and to explain their reasoning.

Step 7
Pupils are given a list of key vocabulary (before, after, earlier, later) and find different ways of expressing the same relationship, for example: 'X happened before Y', or 'Y happened after X'.

Extension activities
- Ask pupils to construct their own timeline on which to place the events.
- Use a list of events that span both AD and BC.

Activity 4

Using generations

Key concepts
Time is a difficult and impersonal concept for children to grasp and one method of making it meaningful is to relate it to the family context. In this activity the concept of generations is explored and related to the individual. A generation unit is usually taken to be 25 years, and on the whole children will recognise the number of ancestors to the grandparent level.

The mathematics
The mathematics used in this activity is based on multiplying numbers, doubling, trebling and so on, and gives practice in multiplication and the generation of number sequences. It also presents a good opportunity for children to gain experience in working with large numbers. Some may need to use a calculator for this and can be challenged to carry on calculating for larger numbers than the calculator can handle. This gives children the chance to use the language of number and extends their use of doubling, halving, trebling and

Numeracy and Mathematics Across the Primary Curriculum

dividing. Higher attainers could be asked to tackle the rate of growth of different sequences and to draw graphs to show their non-linear growth pattern. You can also use the activity to develop children's conceptual understanding of measuring by considering 'the generation' as an alternative standard unit for measuring time.

Step 1
The pupils make a simple matrix with columns headed 'Generation' and 'Number of ancestors'.

Step 2
They explore the concept of how many grandchildren and great grandchildren a couple could have based on different assumptions about how many children there are in each family.

Step 3
They decide on a rule for working forward through the generations of children (e.g. doubling, trebling).

Step 4
They complete the matrix using either multiplication or a calculator until the twentieth generation, verbalising the numbers in each total.

Generation	Number of ancestors
1	2
2	4
3	8
4	16
6	64
8	256

Table of results

Extension activities
- The activity above demonstrates the powers of number, in this case related to the number two. The first generation is 2 to the power of 1, the second generation is 2 to the power of 2 ($2 \times 2 = 4$), the ninth generation is 2 to the power of 9 ($2 \times 2 \times 2 \times 2 \times 2 \times 2 \times 2 \times 2 \times 2 = 512$). These can be inserted in the table in another column.
- Form a column with the dates at 25-year intervals from the birth year of the individual. If a child was born in 1995 then notionally the generation before was born in 1970, and the third generation in 1955. When was the twentieth generation born? How many direct ancestors could the child have in that year?

- A further activity might involve the idea of parents, grandparents, great-grandparents, great-great grandparents and the pupils might provide a formula for working out how many 'greats' there would be for identified generations: the third generation is one 'great', the fifth generation is three 'greats', and the twentieth generation will be $20 - 2 = 18$ 'greats'.
- Challenge pupils to find out whether or not the table works the other way around. How many descendants would someone have if they were born 20 generations ago?
- The data in the table might be recorded as a personal descendant tree with two parents, branching to four grandparents, and so on.

Activity 5

Buildings and symmetry: Why were symmetrical shapes used in the local church?

Key concepts
Buildings contain a great deal of mathematics due to the nature of their plans and elevations. The maths has been used to ensure the structure can stand up, has useful space for a particular function and looks attractive. Identifying the mathematics and its function offers a number of opportunities for insights into particular eras and comparison between periods.

The mathematics
Symmetry can help to identify the motives of designers and builders and gives clues to the function of a structure built in the past. In identifying symmetry (rotational or reflective) in a structure – either in plan or elevation – pupils will be describing 2D and 3D shapes and drawing them with increasing accuracy, recognising geometrical features and properties.

Step 1
Ask the children to identify the types of symmetry and their position by drawing sketches of the symmetrical elements of the building.

Step 2
See if they can infer the possible use of symmetry in the design or function of the structure.

Step 3
Their results can be kept in a table.

Sketch of the symmetrical figure	To take equal weight	To get equal light or heat on all sides	To give equal room	For decoration
△ a window	yes	yes	no	no

Table of results

Step 4
Convert results into a Venn diagram using four sets.

Step 5
Identify symmetry used for more than one purpose.

Extension activities
- How many axes of line or rotational symmetry can you find? You can mark these on a photo in different colours.
- What would have to be changed to make the building completely symmetrical?
- Help the children to look for examples of the transformation of shapes: enlargement, translation, rotation, reflection.
- Apply this activity to artefacts. Many of these have symmetrical properties that are used for aesthetic reasons, and to ensure balance and use from all sides, or to guarantee that equal amounts of substances can be held. Ask the pupils to consider how the designs reflect the need for equal access when the object is turned.

Activity 6

Historical interpretations: How are these records similar or different?

Key concepts
Interpretations of a particular event, person or place can vary according to a number of factors: they might depend on who made them, when they were made, the amount of information available at the time, the purpose for which they were made, and the type of presentation used.

History: Mathematics and evidence from the past

The mathematics
Often interpretations use mathematical concepts to illustrate an aspect of the past with a range of formats related to different purposes. For example, a bar chart of population in 1860 is an interpretation of history, although many of us might find this a strange thought. Other mathematical formats frequently used in communicating interpretations of the past include:

- classification (topic web, Venn diagram, matrix)
- ordering in time or type (timeline, list, flow chart)
- showing position in space (map, plan, scale drawing, scale model)
- showing similarities with other pieces of evidence (table, graph, histogram)

It is important that children are made aware of these various formats.

Detecting similarities and differences between interpretations enables pupils to ascertain why variations occur, and using a Venn diagram can be a valuable aid to analysis. The overlapping set will contain the similarities, and the discrete sets will contain the differences. A large overlapping set is likely to mean that the same source material has been used to make both interpretations, or that one has been used as a base for the other. If, on the other hand, the overlapping set is small, then different sources may well have been used or one might contain new information not available when the other was being constructed. The activity requires the pupils to examine two interpretations and to detect differences and similarities between them.

Step 1
The pupils choose, or are given, two different interpretations of Boudicca's Revolt.

Step 2
They examine both of these, looking for elements that are similar or different and noting them down.

Step 3
They transfer their findings to a Venn diagram.

Step 4
They examine the diagram and decide whether the interpretations are similar or different.

Step 5
They try to explain their conclusions.

Extension activities
- Use documentary sources alongside pictorial sources, models or similar reconstructions.
- Encourage children to compare original sources, such as artefacts or buildings, with interpretations to see what has been added to, or left out of, the original source.
- Construct a timeline with the pupils to record how interpretations or representations of an event, place or person have changed when new information became available.

3 Geography
A place for number

Geography is the study of people and places and the interactions between them. It is primarily an enquiry-based subject that involves systematic analysis of places, at a range of scales, in order to explore how natural and human processes interact to produce spatial patterns that show similarity and difference.

The activities in this chapter are designed to make children aware of the characteristics of places and to scaffold the geographical understanding with mathematical concepts and skills in order to make these characteristics more accessible.

Where is the mathematics in geography?

Mathematics is widely used in geography as part of what has been termed 'graphicacy' – non-verbal and largely pictorial ways of showing information about place: maps, plans and diagrams of various sorts. Mathematical concepts are also used in location, for example using coordinates, in direction where children employ the language of rotation, and in distance where they use a variety of measures. Number is widely used as a means of quantifying places and while many of the calculations may not be very difficult, any errors of miscalculation will be easy to identify as the answers have meaning within a geographical context.

Activity 1

Signpost mapping

Key concepts
Defining the position of a locality in relation to others is an important concept that needs to be grasped if children are to understand the nature of place.

The mathematics
This activity uses the concepts of distance, measurement and angles, which gives pupils the opportunity to apply and extend their measuring skills in a purposeful context. It can also be developed to incorporate scale.

Numeracy and Mathematics Across the Primary Curriculum

Step 1
Set the challenge: The school needs to make a signpost for parents and visitors that has directions and distances on it.

Step 2
A group chooses a location near the school entrance.

Step 3
They put a point in the middle of their paper to show the location, then put the clipboard and the paper on the ground, leaving it unmoved until the end of the activity.

Step 4
The pupils identify another location that is relevant to parents or visitors, for example the Secretary's office, and point to it.

Step 5
They stand on the opposite side of the paper from the new location and draw a line from the central point in the direction of the place they have chosen.

Step 6
They then pace or measure the distance to the chosen part of the school and write this distance on the line, recording whether it is in paces or metres.

Step 7
The pupils decide whether to work clockwise or anti-clockwise from the first location and then repeat the process for five other parts of the school building.

Step 8
The signpost map is titled and used by others to see if it is effective.

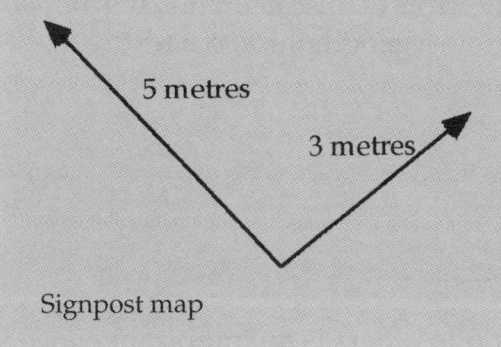

Signpost map

Extension activities
- The measurement aspect can be developed from informal measures (foot length, pace) to more formal examples, such as 'to the nearest metre', or from '15 m and 5 cm' to '15.05 m'.

- The 'signposts' can be drawn to scale in order to represent the distance more visually.
- The angles can be measured from the drawing, and both distance and angles can be included in a set of written instructions for another group to make the signpost.
- A similar activity, using an overhead projector (OHP) and commercial maps or aerial photos of varying scales, can be used to relate the school to its wider locality. Taking the school as the central point, the actual distances can be measured using the map scale and differences between straight line and possible road or rail routes can be measured and compared.

Activity 2

Developing trails

Key concepts
Trails are a fun and adaptable way to explore localities. They provide a cross section through places. Pupils can differentiate between trails by guiding their audience over different distances and in varying directions, altering the starting point or the number of 'stops' along the way. The concepts used will also differ depending on who is making the trail and who will be using it. It might, for example, be a settlement trail where the focus is largely on environmental features. In all cases the language and level of the concepts will be differentiated by the need to engage a particular audience.

The mathematics
The level of directional language can vary from the use of straightforward everyday language (e.g. left, right) to the use of polar coordinates. Children should be encouraged to notice the increased precision that can be obtained with more sophisticated techniques. Besides using mathematics for direction, the foci for each stop could encourage use of number, shape and so on.

> *Step 1*
> The pupils identify their audience and consider the level and types of directional information they want to give ('Go to the old tree', 'turn left', 'turn north', 'walk on a bearing of 350 degrees') and the distance information that would be most useful (paces, metres and so on).
>
> *Step 2*
> Depending on the level of directional information chosen, they identify a range of significant places in the school. These might be places that have different weather conditions (inside and outside), places with contrasting views or different functions, and those with varying associations for groups of children.

> *Step 3*
> They visit each place and make notes on what they will need to include in their instructions (their 'scripts' will eventually be recorded onto an audiotape). Children can use an attributes matrix to record their observations with stops on the horizontal axis and the attributes of the places on the vertical axis. Language used could vary from 'marshy' or 'wooded' to more specific geographical vocabulary like 'detached' or 'semi-detached'.
>
> *Step 4*
> They choose a starting position and once they have decided how they will give instructions, they write the script, trial it for the audience and make adjustments before recording the text for others to use.

Extension activities
- The directions can be given in cardinal or ordinal points of the compass, or in degrees by using a 360 degree protractor or a prismatic compass.
- Consider a range of audiences – parents, governors, tourists, foreign visitors.
- Produce a flow chart of instructions – including decision boxes – for making a tape.
- A supermarket, where shelves are parallel and a plan of their layout is usually available on request, is a good location in which to use the language of right-angled direction and distance. Pupils can get a recipe card from a supermarket and plot the paths to shelves that contain the necessary ingredients. Ask them to plot the most economical route in terms of distance or use an overlay grid to give coordinates of products.

Activity 3

The weather

Key concepts
The weather is the most significant factor contributing to change in places. Although the dynamics of atmospheric change are complex, the resulting consequences, temperature, wind speed and direction, cloud cover and rainfall are easily identified and recorded by young children. However, rather than try to teach these factors as 'weather' it is perhaps better to refer to them in terms of 'climate', as they are then seen as qualities of a place rather than something that is *happening* to a place.

The mathematics
Mathematics can help to quantify many aspects of weather and to allow a greater understanding of the processes of wind – precipitation in particular.

Weather recording uses a range of instruments and scales. The scales can be non-standard (e.g. described in terms of types of clothes needed, or using words such as 'calm', 'breezy' or 'windy') or formal (using wind speeds, temperature and wind direction). A bar chart will be used to compare and show similarities and differences. Children should be encouraged to notice the discrepancies between measurements that are made by counting (discrete data) and those taken with measuring instruments (standard or non-standard), which produce fractional values (continuous data).

Step 1
Visit a number of locations around the school, which have been chosen because of their different attributes (e.g. sheltered, open, high, low).

Step 2
A number of groups of pupils simultaneously record aspects of the weather using the same scales and similar instruments. They record wind speed, wind direction and temperature.

Step 3
A set of graphs can then be made. Ask one group to draw a graph showing the temperature at each place; a second group could draw one to show the wind speeds; a third group could show wind direction by drawing a wind rose – a representation that uses the compass and the length of any one directional 'arm' to show the scale of wind speed.

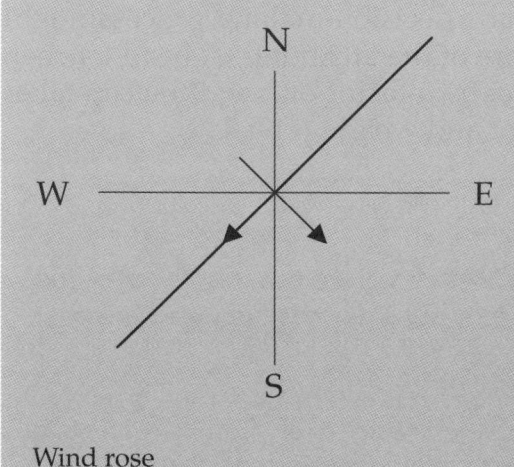

Wind rose

Extension activities
- Pupils choose a location in the school area and record the temperature every 15 minutes throughout the school day. They plot the temperature points on a graph with temperatures on the vertical axis and time on the horizontal axis. The points are joined with the line of best fit and pupils then make some statements about the data they have collected.

- Measure micro-climates around the school, such as wind funnelling, and shade or rain shadow caused by structures or trees.
- Link via email with other schools in a variety of locations and get them to undertake the activity at the same time. Atlases, the Internet, or other forms of information storage can provide data. Pupils might work out local time for schools outside the UK.

Activity 4

Improving the environment: Think globally, act locally

Key concepts
An important aspect of environmental change is the ability of the individual to improve places. Changing environments to make them more pleasant often involves the addition of facilities, and the ability to make changes in this way can empower children who may feel helpless to deal with more global issues.

The mathematics
This activity follows the decision-making process behind the placing of a bench (donated by the PTA) in the playground. The mathematical tool used here is an attributes matrix. In this case the best position for the bench would have three main attributes: it would be sheltered from the sun, be visible from all points in the playground for safety reasons, and travel to it must not damage the grass or produce mud that might be carried into the school building. A rating system of 1 to 10 is employed. Those involved in the decision-making process might feel that it is desirable to weight one or more of the attributes. A head teacher, for example, might weight the safety aspect by a factor of 2 and the caretaker might consider the mud problem significant and weight it by 3.

Step 1
Having decided on the focus of the activity, 'Where will we put the bench?', the pupils brainstorm the sorts of attributes that they consider necessary for the best positioning of the bench.

Step 2
They decide if any of these attributes need weighting. In this case the sheltered aspect is considered most significant and is weighted by 2.

Step 3
They choose a range of places that they will quantify in terms of suitability, using the attributes as a guide.

Step 4
They undertake the quantification exercise in each place using informal measures.

Geography: A place for number

Step 5
Results are then recorded in a table so comparisons can be made (scored out of 10). A map can be used alongside the table, to show exactly where the places are.

	Sheltered (×2)	Good view from playground	Not too muddy	Total
Place A	8 × 2 = 16	4	5	25
Place B	4 × 2 = 8	6	8	22
Place C	5 × 2 = 10	8	2	20
Place D	9 × 2 = 18	8	6	32

Table of results

Step 6
Pupils interpret results and decide that Place D would be the best position for the bench.

Extension activities
- A preference matrix could be used to bring together the results from different groups' decision making. To use a preference matrix each group should order the different locations according to preference (this may well match the results of their attributes matrix). If the ranking scores the most desirable as one, then the winning item is the one that scores the lowest.
- Positioning bird boxes or bird feeders will require research into the types of environment particular species prefer. It will also involve decisions about views from classrooms.
- Construct the attributes matrix as a spreadsheet so that the effect of choosing different weightings can be explored easily.

Activity 5

Quantifying environmental threats

Key concepts
Environments are subject to threat from a variety of human and natural agents. These threats can be quantified through analysis and, if action is taken to reduce the damage through remedial measures, the improvement might also be quantified.

The mathematics
This activity explores the effects of environmental change and offers a means for children to gauge informally the influence of factors on an environment in terms of threat. The chosen tool is a ripple diagram – a series of concentric

circles that allow the children to place threats to an environment on a scale of distance from the centre of the diagram. The further away from the centre the smaller the threat. Children can use non-standard measures (e.g. rating from one to ten) to quantify each threat, in order to decide where they should be positioned on the diagram.

Step 1
Pupils identify the threats to an environment through examining a description of the dangers or using a photograph.

Step 2
They write these on separate cards.

Step 3
They position these cards in the appropriate place on the ripple diagram. Severe threats come closest to the centre of the diagram, less severe threats in the middle area and the least threatening towards the outside.

Step 4
The pupils analyse the diagram and identify the most serious threats.

Step 5
The pupils suggest remedial measures to deal with the threats.

Extension activities
- To refine their thinking, pupils construct an attributes matrix (Figure 3.1) in which possible damage to the environment is weighted according to seriousness. For example, the effect on the view might be weighted by two.

	Poor air quality	View spoilt ($\times 2$)	Over crowding	Litter	Damage to landscape
Hotel					
Car park					
Quarry					

Figure 3.1 An attributes matrix comparing levels of environmental damage

- An attributes matrix is also valuable for checking whether improvements have been successful. Children have the opportunity to analyse the case before and after action has been taken.
- Use photographs taken in the past to enable children to explore the damage done to a place by development or 'improvements'.

4 English
Patterns of language and patterns in language

Current initiatives in the teaching of literacy recognise the significance of four main areas of communication:

- *Writing* – to communicate ideas or information to someone else or to write for oneself to read at a later date.
- *Reading* – to make meaning from other people's writing.
- *Speaking* – to communicate ideas or information to someone else orally or to record oneself speaking so that it can be listened to at a later date.
- *Listening* – to make meaning from other people's speech.

The acts of speaking, listening, reading and writing are in many ways symmetrical. Speaking and writing *create meaning for somebody else*: In creating meaning there is usually an audience and a purpose that is represented by a specific structure. Listening and reading *recover meaning from somebody else*: In receiving meaning someone is part of an audience and has to determine the purpose represented by the specific structure.

Knowledge base of the subject

Although the knowledge base of the subject includes grammar and punctuation, our present study is primarily interested in the main formats for communication that have been highlighted as most suitable for a Key Stage 2 mathematics focus. These structures or processes can be identified as:

- Recounting
- Reporting
- Explanation
- Instruction
- Persuasion
- Discussion

They are generally apparent throughout *non-fiction* writing, reading, speaking and listening and feature prominently in some *fictional* narratives where recounting may be combined with elements of reporting, explanation and discussion. In some of these cases, the narrative may have been deliberately set up as a series of fictional instructions. Recounting and persuasion also feature extensively in poetry and authors often use instruction to impart information, particularly in modern children's verse.

Where is the mathematics?

Mathematics offers us a huge range of vocabulary to describe things and to help us translate anything we don't understand into more easily recognisable terms. These are known in mathematics as nominal scales and include shape and numbers.

When writing fiction and non-fiction, children are involved in problem solving and as part of that process they can use mathematics to help them represent and develop their understanding. They swap between different modes of communication, in this case translating verbal or symbolic language formats into pictorial ones, thus using alternative representations to help them clarify what they are hearing, saying or writing. Such representations can be used to:

- clarify their own thinking before writing and speaking;
- clarify what others are saying or writing;
- make representations for others to change into a verbal format.

The mathematics enters the process at the point at which ideas and content matter are generated and, perhaps more powerfully, in the organising and structuring of the text framework (see Table 4.1).

Table 4.1 Using mathematical formats to communicate meaning in writing

Purpose	Mathematical formats for communicating meaning
Describing	Nominal scales
Classifying	Topic web, Venn diagram, tree diagram, attributes matrix
Comparing	Bar chart, rating, attributes matrix, preference matrix
Ordering	Time chart, ordered list, network, flow chart, Gantt chart
Positioning	Scale model, map, scale drawing, plan
Interrelating	Table, histogram, scatter graph, graph

Where is the mathematics in poetry?

Understanding the meaning of the language in a poem makes use of the same skills and techniques as other texts, but there are additional features such as aspects of the physical layout and the structure of the text, which emphasise pattern and add to the meaning. This leads to the use of counting, shape and algebraic thinking in communicating meaning.

Nearly all poems make some use of rhythm. In order to appreciate this, the reader must scan the poem, identifying the number of syllables in a line focusing in particular on the number of those that are stressed. This pattern of stressed syllables defines the metre or rhythm of the poem. Thus, counting, usually unconsciously, plays a key role in understanding, reading and writing poetry.

Some poetry makes explicit use of shape to enhance the meaning of the text by selecting an image that symbolises the content of the poem. In this case, the basic mathematical shapes must be understood and then used in a symbolic way so that their mathematical properties are emphasised, thus creating the potential for pupils to develop algebraic thinking in using a symbol to carry meaning.

The use of pattern is a key feature of mathematics and identifying the patterns in poetry algebraically can enhance the appreciation of the meaning and technical skills embodied in the poem. The identification of a poem's pattern is the start of the process of generalisation. By stressing some features and ignoring others, the pattern can be extended to identify the significant features of a type of poem. This pattern may be visible in the line length and metre, or in the rhyming structure. Again much of this structure may only be appreciated unconsciously, but if some attempt is made to record this pattern, either in words or symbolically, then this is the beginning of algebraic thinking.

Activity 1

Fiction: The Giant's jigsaw

Key concepts
There are many forms of *narrative* text, which usually tell imaginary stories with the purpose of entertaining or engaging the reader in a fanciful experience, attempt to explain a phenomenon (myths and legends), or convey a message (fables). Writing narrative has four main stages: orientation, events, complication, and resolution.

The mathematics
In the opening *orientation* stage, setting, time and main characters are established. The most useful format for organising the attributes of main characters will be topic webs. Tree diagrams will help children interpret the action in relation to other starting events and the physical location in which the action will take place can be developed through scale models, maps, scale drawings and plans. *Events* are decided in the next stage; timelines, ordered lists, flow charts, calendars and Gantt charts are useful for putting the actions in sequences either as a single thread or as parallel happenings. An element of conflict is introduced in the *complication* stage, before leading ultimately to a *resolution* of some sort. This might be achieved through the use of a preference matrix or attributes matrix, which will help quantify the possible endings.

The Giant's jigsaw

The Giant loved doing jigsaws. He always made himself comfortable by dressing in old denim jeans, a t-shirt and soft leather shoes. The Giant always did a puzzle in the same way. First, he made the border with its straight-edged pieces. Then he did all the people. Then he concentrated on the 'fiddly' bits – with lots of clouds, or sea, or curtains – where it takes a long time because all the pieces look the same. He was good at finishing jigsaws and he always put the pieces away carefully. He hated doing a jigsaw that had pieces missing. If he had thought that someone had deliberately taken a piece, he would have been very angry. However, he knew that if any of the pieces were lost it must be his own fault for he was very shy and lived on his own. Often this made him feel very sad and lonely. This was also the reason why he did so many jigsaws: it helped to pass the time.

Step 1
Children use a topic web to record the characteristics of the person they are hearing about or in planning a similar character for their own fictional writing.

Step 2
They use a timeline to plot/summarise the action, or record information in a table. This will ensure that various characters can be compared and that consistency is maintained in their writing.

Activity 2

Non-fiction: The Titanic

Key concepts
The 'recount' genre involves the retelling of events in order to convey information or for purposes of entertainment. It includes formats such as diary, autobiography, news report, letters, journals, descriptions of events or visits, and anecdotes. Examples of all of these formats should contain a setting and events.

The mathematics
Again topic webs will be the most useful format for organising the attributes of the main characters, while tree diagrams will help readers to interpret the action in relation to other starting events. The physical place in which the actions take place can once again be developed through scale models, maps, scale drawings and plans. When children reach the stage at which *events* are chosen, timelines, ordered lists, flow charts, calendars and Gantt charts will be useful for putting the actions in sequences either as a single thread or as parallel happenings.

At 11.40 p.m., lookouts Frederick Fleet and Reginald Lee were up in the crow's nest. Suddenly, Fleet was alarmed by a haze he saw on the horizon. He rang the warning bell three times and telephoned the bridge, shouting, 'Iceberg, right ahead!' As the Titanic steamed directly towards a huge black mass of ice, First Officer William Murdoch turned hard to port, ordered the engines to stop and for the 15 watertight doors to be secured. Despite these efforts, the Titanic scraped against the iceberg for 10 seconds on her starboard side.

After the impact, Captain Smith rushed into the chart room and officers sent for the ship's chief engineer, Thomas Andrews. Fourth Officer Boxhall made an initial inspection of the forward areas and reported that he saw no damage. Captain Smith and Andrews then inspected the ship for themselves and found that the iceberg had damaged the plates in five or six of Titanic's watertight compartments. They quickly realised that the ship would sink, since she could not remain afloat with more than four compartments flooded. Seawater had already risen 14 feet above the keel, and water was bursting in from the forepeak to boiler room five as the ship began sinking by the bow. Many of the passengers, however, had no idea that the Titanic was in grave danger.

After the collision, many gathered in the corridors half-dressed. 'Everyone seemed confident that the ship was all right,' recalled first-class passenger Henry Sleeper Harper. At last, however, the call came for all passengers to come up on deck wearing their life belts, and soon after midnight, Captain Smith directed crew members to ready the 16 wooden lifeboats and four collapsible boats. The noise on deck was horrendous as steam was released to ease pressure on the Titanic's boilers. Over the din, Officer Lightoller shouted to the Captain for permission to begin loading the boats, and the Captain nodded in agreement.

On the port side only women and children were, for the most part, permitted to climb aboard the lifeboats, while on the starboard side men were allowed to get into the boats if no women were in sight. Still, many women were reluctant to leave their husbands and the apparent safety of the huge liner for a 70-foot drop down to the dark ocean in the tiny wooden boats. Some had to be forcefully picked up and dropped into the lifeboats by crew members. Very few of the boats were loaded to their capacity with passengers.

At 12.45 a.m., Quartermaster George Rowe fired distress rockets as lifeboat seven was lowered with only 28 people aboard, even though it could have carried 65. Realising the danger of their situation, many third-class passengers gathered in prayer, and five men jumped into lifeboat five as it descended, seriously injuring a female passenger. By 1.15, the Titanic's bow had plunged beneath the surface. Even as water was rising in the ship, the band continued to play and the gymnasium instructor was assisting passengers on the mechanical exercise equipment.

At 2.05 a.m., after the last wooden lifeboats and two of the collapsible boats had gone, Captain Smith told the crew to look to their own safety. A group of men on deck struggled to release one of the two remaining collapsibles, which was lashed to the top of the deckhouse. Hundreds of other passengers were praying, crying, and jumping from the ship. At 2.10, the bow dropped further, sending water up the deck as passengers struggled toward the stern. Seven minutes later the stern rose almost vertically into the sky, stopped and pivoted. The Titanic's lights flickered and finally went out as the ship at last eased down into the water and sank below the surface of the sea.

Step 1
Pupils analyse the text recounting the events surrounding the sinking of the Titanic.

Step 2
They make an ordered list. This allows them to make sense of a complex series of actions and identify the relationship between a number of key 'players', i.e. the ship, her crew, the passengers and the iceberg.

Time	The iceberg	The crew	The passengers	The ship
11.40 p.m.	Is sighted Scrapes against ship	Sight iceberg Change of course ordered and safety measures taken Chief Engineer inspected forward areas	Have no idea of danger	Turned hard to port Engines stopped Watertight doors secured Scraped against iceberg Damage to watertight compartments Seawater bursting in
After Midnight		Captain orders crew to make ready lifeboats Orders boats to be loaded	In corridors half dressed On deck in life jackets Port side women and children into boats and on starboard side men into boats	Boilers release steam
12.45 a.m.		Distress rockets fired	Passengers praying Five men jump into lifeboat injuring woman	
1.15 a.m.		Band continues to play and gym instructor assists passengers on exercise equipment		Bow plunges beneath surface
2.05 a.m.		Crew look for own safety	Last lifeboats gone. Hundreds of passengers pray, cry and jump	
2.10 a.m.			Passengers struggle for stern	Bow drops further and water rises up deck
2.17 a.m.				Stern vertical in sky, stops and pivots. Lights flicker and ship sinks

Ordered list

Activity 3

Alligators and crocodiles

Key concepts
The 'report' genre describes the way things are. It includes formats such as encyclopaedias, information leaflets or posters, dictionaries, indexes, reference texts, thesauruses, non-chronological reports, glossaries and directories. Report writing should include an opening generalisation or classification, followed by description and ending with a summary.

The mathematics
The *opening generalisation* or *classification* can be planned using a topic web, Venn diagram, tree diagram or attributes matrix, all of which will demonstrate the relationship between the various elements. In the *description* stage, nominal scales, shapes, scale models, maps, scale drawings and plans can be valuable for quantifying aspects of the focus.

To help design a *summary* that brings everything together, 'collecting' formats such as tables, histograms, preference matrices, attributes matrices and bar charts are used.

Alligators and crocodiles

The American alligator (Alligator mississippiensis) and the American crocodile (Crocodylus acutus) are both reptiles. They have elongated, armoured, lizard-like bodies with muscular flat tails, and long snouts with nostrils at the end that allow them to breathe when their bodies are submerged. They also have four legs with five toes on the front feet and four on the rear. The skin on their backs is armoured with rows of embedded bony plates called scutes. Adult crocodiles and alligators grow to a length of around 13 feet.

Young alligators have bright yellow stripes and blotches, and the adults are dark with pale undersides. Their snouts are rounded and shovel-shaped. Adult crocodiles are grey-green across the back with pale undersides and the young have dark cross-bands on their backs and tails. Crocodile snouts are sharp and pointy and they have a fourth tooth on the lower jaw, which is exposed when the jaws are shut.

Both species eat crabs, fish, birds, turtles, snakes and mammals. Alligators prefer to live in large shallow lakes, marshes, ponds, swamps, rivers, creeks and canals. Crocodiles choose estuarine areas along coastal mangrove swamps. They build their soil nests in the form of mounds or holes, while alligators' nests are mounds of vegetation.

> *Step 1*
> Ask the children to analyse the text on alligators and crocodiles.
>
> *Step 2*
> They should then use the information given in the extract to draw a Venn diagram, showing the similarities between both types of reptile in an overlapping set, while illustrating strongly the difference in the discrete sets.
>
> *Step 3*
> The pupils then classify the reptiles using appropriate sorting questions and design a tree diagram to illustrate the attributes of the creatures.
>
> *Step 4*
> Finally, the data are arranged in a table with the attributes of length, colour and similar characteristics at the top, which can be used by others to write a report.

Activity 4

A chip shop for Charlbury?

Key concepts
'Persuasion' genres promote a point of view. They include formats such as public information posters, advertising leaflets, magazine articles and editorials. Like report writing, composition in this category of written work must include three stages: thesis or overview, argument put forward with supporting evidence, and summary, where the opening position is restated.

The mathematics
The *thesis* can be designed using a topic web to show the relationship between facts or opinions. *Argument* aspects are best explored using formats that allow a great deal of detail to be portrayed: bar chart, rating, attributes matrix, preference matrix, scale model, map, scale drawing, plan, timeline, ordered list, flow chart, calendar or Gantt chart. Mathematical formats that show the many aspects of the issue clearly are best employed in the *summary* section. These might include a topic web, Venn diagram, tree diagram or attributes matrix.

A Chip shop for Charlbury?

The local community is divided over the plans to build a chip shop in this wonderful Cotswold town. Should the plans be ditched? No!

Firstly, a fast food outlet like a chip shop will bring opportunities for part-time employment for some young people in an area where there are not many jobs of this sort.

Secondly, the chip shop will provide a service that is lacking in the area. The nearest outlet of this type is twelve miles away and the only other sources of meals of this sort are in the town's pubs. Over two-thirds of the 100 people we interviewed were in favour of the chip shop: 68 were for and 32 were against.

Although some might say that the smell from the cooking process is aroma pollution, the smell of fish and chips is traditionally English. As the shop will be in a conservation area, the neon sign that will advertise the chip shop's presence will be carefully designed to minimise light pollution. Furthermore, the owners will be especially careful to ensure litter is kept to a minimum through the provision of litter bins. The chip shop will not disturb the quiet atmosphere of the town; despite the fact that it will attract people from a wide area, the road is far too narrow at this point to encourage young people to form groups.

The chip shop should be allowed to go ahead. Not only will it be an asset to the local area, it will also provide jobs and encourage people from outside Charlbury to visit the town and spend money in other shops. It will definitely be an important feature of the town.

Step 1
The children consider the scenario and think about how to apply mathematics to back up their arguments.

Step 2
They draw a pie chart to represent the responses to the chip shop survey and make a plan of the location of the shop so that their points can be made more forcefully.

Step 3
They then write an attributes matrix showing 'before' and 'after', which can be used to back up their arguments.

Activity 5

Poetry: Patterned poems

Key concepts
Strongly patterned poems provide a good source of models on which children can base their own writing. In this activity, the task for the children will be to identify the pattern in the examples they are shown. They will then create their own poems, in a similar form to those they have seen, in order to express an idea or feeling. In poems of this sort, the patterns are often based on the number of syllables in a line and various forms have been developed from the original Japanese *haiku*. As well as being well known for its striking syllabic arrangement, this type of poetry is also often strongly associated with particular topics

or themes to the extent that meaning is significant in establishing the form. For example, a haiku usually evokes a singular aspect of nature, with the first line setting the scene, the second line introducing an action and the third line bringing them together. To read or write this form of verse one must be able to understand, develop and communicate a novel perspective on nature. The form itself can be interpreted more or less precisely; some writers regard the *haiku* structure as unbreakable while others will allow variations if the meaning to be communicated is better served through flexibility. Other forms that can be explored, roughly in order of increasing complexity, include the *tanka,* the *cinquain* and the *diamonte* (for further explanation of these terms see Mallett 2002).

The mathematics
Mathematics has a significant part to play when it comes to identifying the structure of these poems. In order to recognise that a poem's structure is based on the number of syllables in each line, children will only need basic counting skills and the ability (i.e. the reading skills) to identify syllables rather than words. Analysing the arrangement of syllables in poems may be helpful in developing children's spelling. The identification of the pattern in meaning, however, is more complex and requires thinking that is laying the foundations for algebra. To construct a template for the type of poem being studied, the pupils must stress some features that are common to each example (classifying meaning and the type of words used) while ignoring the features that are different (the topics covered and the actual words used). This is strongly related to the mathematical thinking required for generalising in algebra. For example, one form of the diamonte structure follows this pattern:

Line 1: **one word (noun)**	Means the opposite of the noun in Line 7
Line 2: **two words (adjectives)**	Describe Line 1
Line 3: **three words (verbs ending in 'ing')**	Describe Line 1
Line 4: **four words (nouns)**	The first two refer to Line 1, the second two to Line 7
Line 5: **three words (verbs ending in 'ing')**	Describe Line 7
Line 6: **two words (adjectives)**	Describe Line 7
Line 7: **one word (noun)**	Antonym to Line 1

Notice that in this form it is words that are being counted not the syllables. In this type of poem the diamond shape and the symmetry of the pattern are being used to emphasise the transition between the two contrasting nouns. 'Snow/beach' might be chosen for Lines 1 and 7 for example. The patterns involved are complex and will not be captured by any straightforward representation. Although the example above does not immediately look like an algebraic form, it does have the key elements of using symbols to represent some aspect of generality. The noun in Line 1 acts as a variable that holds a place for any noun that could be located there; the idea that it must be the opposite of the noun in Line 7 is placing a constraint on the nouns that can be chosen.

Dark
Thick, scary
Clinging, blinding, freezing
Night, cave, day, beach
Sparkling, smoothing, warming
Clear, friendly
Light

> *Step 1*
> Present the children with a number of examples of haiku and read them together.
>
> *Step 2*
> Focus on one of the poems and challenge the children to identify the events that are described.
>
> *Step 3*
> Identify the feelings that the poem evokes by asking pairs of children to tell each other what it makes them think of and then collect some examples from the whole class.
>
> *Step 4*
> Ask the children to identify the similarities between examples. Next, encourage them to draw out and clarify the features associated with meaning and those associated with the number of syllables.
>
> *Step 5*
> Taking a poem at a time, the children identify the number of syllables in each line by underlining and numbering the example.
>
> *Step 6*
> Try to get the children to notice (where appropriate) the use of the present tense, the infrequent use of adjectives, simile or metaphor, and the way in which the ideas are conveyed with the minimum number of words.
>
> *Step 7*
> Finish off by asking a number of pupils to read their favourite examples aloud and to say what it makes them think of.

Extension activities
- Prepare a poem for reading aloud. The quality of the reading will depend to some extent on the understanding of the structure of the poem.
- Use cuisenaire rods (coloured underlining or pattern blocks) to represent aspects of the pattern. When some understanding of a form of notation has been reached try rearranging the cuisenaire rods and see if the resulting

pattern could be used as a template for writing a poem. Bear in mind that most of the possible rearrangements will be difficult to use.
- By identifying and copying the pattern of meaning and language in a haiku (or other type of poem) pupils can write their own poems using examples as a template and building up a general description of the type of poem. They may need some guidance on the subject matter to choose; with haiku the seasonal theme may make it easier for children to find a starting point. Successive redrafting will help them to refine their initial ideas and gradually match the template.

Activity 6

Reading and writing poems that rhyme

Key concepts
Poems that make use of rhyme probably do so to gain attention by creating a satisfying melodic effect. In lengthy ballads, originally intended for the oral retelling of a narrative, rhyme may have been used to make the story easier to remember and less prone to random change. When reading poems, the rhythm and the rhyme need to be identified so that the author's intended melodic effect is recreated. This structuring gives the author more control over the way the text is read than in ordinary prose, where the effect is difficult to achieve. When writing a poem, the author must consider what effect the chosen rhythm and rhyming pattern are likely to have and ensure that it matches the content of the story being told. Classic examples use the rhyme and rhythm to give the impression of speeding trains, galloping horses or rushing water.

The mathematics
In order to identify the rhythm of a poem, children will need to count the number of stressed syllables rather than just the words. This has a great deal in common with the counting of rhythm in music and may involve the children clapping in unison as the poem is read or someone beating time with a percussion instrument. Recognising and keeping to the rhythm as they read a poem is a key to successful reading aloud and such success is likely to improve the recitation of ordinary prose. The identification of rhyme is a complex activity as poets use a variety of styles of pronunciation to achieve a rhyme. Thus the identification of the pattern may give pupils a clue as to how an unusual word is to be pronounced, or how a word is to be pronounced unusually. This identification of rhyme can be a useful spelling strategy if new words pronounced like known words can be spelt in the same way. The rhyming pattern can be identified initially by highlighting the rhyming words using coloured pens. This can be developed so that the alternate-line rhyme scheme of a typical ballad is represented in alphabetical symbols as ABAB, a clear example of an algebraic representation where the letters A and B are variables standing for the rhyming words.

English: Patterns of language and patterns in language

> *Step 1*
> Write a ballad about an incident at school: a report of a school netball match, an occurrence in the playground, a school assembly, etc.
>
> *Step 2*
> Discuss and plan the rhyme scheme to be used and set out the pattern using coloured blocks, shapes, symbols or words to represent the sequence.
>
> *Step 3*
> Plan the main events to be reported – perhaps using a topic web or flow chart to keep track of them – and try to arrange them in a suitable order.
>
> *Step 4*
> Children draft the opening verse and then rework it until it matches the pattern; alternatively they adapt the pattern to match the best version of the verse.
>
> *Step 5*
> Write verses for each of the other incidents.
>
> *Step 6*
> Read the poem aloud and challenge the other pupils to identify the rhyming pattern and represent it using coloured blocks or shapes. Use their feedback to identify any further redrafting that might be beneficial.

Extension activities
- Prepare a choral poem for reading aloud, perhaps in a school assembly. Identify and count the beats in the poem and click fingers or beat time in some way. Decide how many voices should read each section of the poem or which parts should be limited to a single voice. Rehearse until pupils can read well with appropriate emphasis and make use of the rhyme and rhythm.
- Omit the rhyming words from a poem and replace them with a colour code, a shape or represent the pattern in words or symbols; the children can then be challenged to find words to fill the gaps.
- Identify the rhyming pattern in a poem or set of poems and find a way of recording this pattern using different coloured underlining, pattern blocks, words or symbols. For example, identify the rhyme and rhythm in clerihews and/or limericks and get pupils to read their favourites aloud.
- Present the pupils with a number of examples of poems and a rhyming pattern represented in coloured blocks, words or single letters. Challenge the pupils to identify which poem has the rhyming pattern you have given them.

5 Physical Education
Building geometrical thinking into playing games

Physical Education (PE) is concerned with the development of pupils' physical skills and the cognitive growth that provides the basis both for understanding and developing skilful performance. The development of physical skills is thus connected to the growth in knowledge of how the body moves and how this contributes to fitness and health; this affects judgements on what constitutes effective performance and ensures that necessary improvements can be made. In the programme of study for PE, the aim to develop cognitive thinking skills is apparent in the range of problem-solving activities that are presented. These include planning and adapting strategies and tactics in competitive games, or creating and composing in the areas of individual performance like dance and gymnastics. The knowledge and understanding developed may have added value in enabling pupils to gain more enjoyment from any sports that they watch.

Knowledge base of the subject

Within PE, the development of the relevant knowledge, skills and understanding is usually thought to occur in six key contexts: dance, games, gymnastics, swimming, athletics and outdoor adventurous activities. The contexts for PE obviously focus attention on the physical aspects of developing pupils' movement skills and knowledge about their bodies' capabilities and limitations. However, there are also significant cognitive components required for pupils to be able to tackle the problem solving within each area. Some argue that PE activities provide the context in which pupils' logical thinking and planning skills first emerge as they tackle the creative and competitive challenges involved. There is also a significant social component to the learning in PE, as pupils learn about themselves, the way they respond to challenges and how they learn to work in teams in both cooperative and competitive situations.

Dance and gymnastics are used as a vehicle for exploring movement. They can help children understand how movement can be characterised, by looking at the ways in which the body can move, the space in which it moves and the way in which the body relates to itself, other people and the environment. In this context, the development of movement skills is often focused on improving precision and accuracy. However, there is also an aesthetic component to movement, which applies to composing and evaluating sequences of movements in gymnastics and to the choreographing and performance of complex dances. The games context provides opportunities to develop physical skills in throwing, catching, kicking and striking while enjoying competitive activities in small groups. The team games involve the development of social skills in organising groups and following the rules of the game. The identification and development of appropriate tactics for attack and defence will also form a significant element of this work. The swimming context allows pupils to learn about movement in a different environment and incorporates floating exercises, swimming using a range of recognised strokes and issues of water safety. The physical skills covered by athletics include running, jumping and throwing in ways that call for increasing precision, speed, power or stamina. In time pupils will learn how to select appropriate strategies to meet their personal challenges or to take part in competitions. The outdoor adventurous activities involve similar physical skills to athletics, focusing particularly on the development of stamina but also including a significant cognitive challenge in following trails, using maps and solving orienteering problems. All of these physical activities can contribute to pupils' knowledge and understanding of fitness and health, and link to the scientific approach to health education.

The significant problem solving within PE occurs when pupils are selecting and applying skills, tactics and compositional ideas in one of the contexts described above, or in cases where they are trying to evaluate and improve their performance. Since these cognitive challenges are related directly to physical activities, this may be an area in which logical thinking first develops or is discussed. Pupils may be able to give rational reasons for their choices and understand alternative points of view because they can relate these readily to a physical situation they have experienced.

Where is the mathematics in PE?

Physical development

The development of physical skills involves work that can be related to shape and space. Symmetry and asymmetry are key mathematical concepts that can be employed to help pupils make sense of the skills they are learning and to enable them to identify what makes performance effective and how it might be improved. The work in dance and gymnastics can be enhanced through the increasingly accurate use of geometrical vocabulary to describe movement. Pupils' kinaesthetic awareness can also develop as a result of this. Work on angle and compass points will be essential components for describing and communicating certain physical skills. Developing and communicating tactics for playing invasion games involves considerable use of geometrical thinking and understanding. Here, children will need to employ mathematical vocabulary to explain and demonstrate actual physical movement in the real world. The ability to represent ideas using models or diagrams would also be beneficial.

For pupils to evaluate their performance they will first need to use mathematics to measure their achievement levels in the relevant area. Much of this will involve the accurate measuring and recording of distances and time, for example the distance they throw a ball or the time they take to run 100 m. This will involve the interpretation of fractions and decimals and children will need to use appropriate skills to order those numbers as they try to compare their performances with themselves and others. Pupils may need to use calculation skills in order to find the speed at which they run or to express their performance as a percentage of school, county or national records. As they try to develop their performance over time, they will need to use data-handling skills and concepts to compare and analyse their own achievements, possibly setting themselves targets for improvement. They might also compare their performance with other pupils in their school or with school, county, national or international records. These extensions may lead to the development of more sophisticated data-handling skills and concepts, including the use of information and communication technology (ICT), as larger data sets are explored.

Activity 1

Passing tactics

Key concepts
In a game of football, what is the best strategy for the attacking pair to make the largest number of passes without the defender touching the ball? How far in front of the running player should the ball be passed?

This activity focuses on the development of passing, throwing and catching skills as well as the appropriate tactics for keeping possession of the ball. As their knowledge of the principles behind the tactics for attacking and defending expands, children should be able to select appropriate times to use the tactic in other competitive games.

The mathematics
The pupils need to learn to send and receive passes while moving. In order to identify the appropriate direction to make a pass the children will need to make judgements about how fast their partners are moving and how far in front of them the ball must be kicked.

This may be estimated in terms of the distance of the ball in front of the receiver, but it would be more useful to identify this using angular measurement, either in fractions of a turn or in degrees. In games where the receiver must catch the ball, acute angles between 10° and 45° are likely to be used. In other games like hockey and football, where the ball may be passed to a *space*, the angle may be larger. Having made their pass the pupils must then think of themselves as potential receivers of the ball and move to find space. This movement will create triangular patterns that have to be repeated at high speed to maintain possession.

The mathematics of shape and space underpinning these ideas can be shown in models that use small-world people or cubes to represent the patterns of

PE: Building geometrical thinking into playing games

movement. Alternatively 'cones' can be set out on the playing area to demonstrate the initial positions and subsequent movements. The more abstract two-dimensional representation of plans or diagrams can be introduced gradually. The practice of playing the game will help pupils to make the connection with the three-dimensional world. On the diagrams or models it will be possible to talk about the angles made between the players and the shapes that the attackers and defenders make as they move. Pupils will also encounter the inherent difficulties in representing a dynamic situation with static diagrams or models.

Step 1
Pupils are told that they will be playing a passing game on a 10 m by 20 m pitch where two pupils will act as the attacking team and a third will be a defender. The attacking team will start with the ball from a designated point at the edge of the pitch. Their challenge will be to make as many passes as possible without the defender touching the ball.

Step 2
The movement expected of the pupils should be shown to them using a physical model in the classroom, e.g. using small-world people, cubes, etc.

Step 3
The angle the players turn to make the pass should be discussed. Draw a diagram to show the triangle pattern that is created by the movement of the passer (A) once the ball has been kicked to the other attacker (B), in readiness for the return pass.

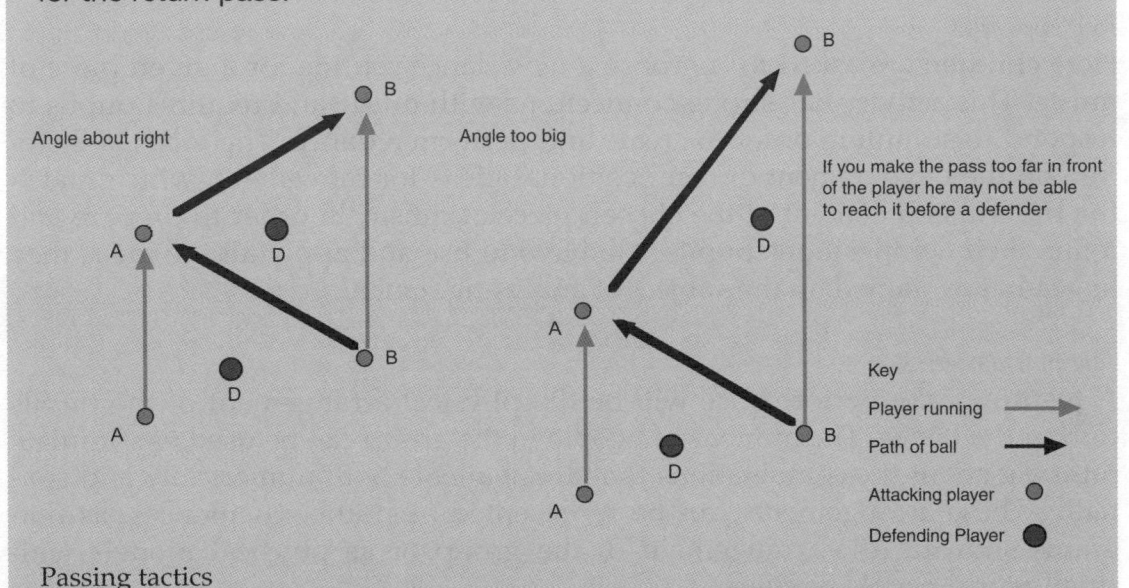

Passing tactics

Step 4
The triangular pattern should be shown on a scale plan of the pitch and on the actual pitch using cones to represent the positions of the attackers.

> *Step 5*
> Let play commence and make sure that pupils swap tasks regularly so that they have experience of attacking and defending.
>
> *Step 6*
> Give the pupils opportunities for recording their performance using diagrams or models. They will then be able to make comparisons with professional players.

Extension activities
- The game can be played with other group sizes, e.g. 3 versus 2, 3 versus 3, or 4 versus 4.
- Goals or targets can be introduced so that scoring points is the objective rather than keeping possession.
- Versions can be played for hockey, football, rugby and basketball, with appropriate variations in the size of the pitch.
- More able or older pupils can plan and rehearse moves from the centre pass or other 'dead ball' situations.

Activity 2

Dance

Key concepts
Here children are asked to compose a new dance routine for a given piece of music. This activity has strong connections with music and requires pupils to respond to stimuli in order to create and perform a dance. The focus is therefore on the development of compositional ideas for movement, which match the rhythm and content of the chosen piece of music. In order to prepare and refine their compositions pupils will have to use and apply all the ideas they have already gained on the subject of analysing dance.

The mathematics
A feature of the performance will be the physical arrangement of the pupils during the dance. The number of people in the group can be used to stimulate thinking about the combinations that are available both numerically and spatially. These arrangements can be represented as number sentences, as diagrams showing the arrangement of the group or as physical models with small-world people or cubes.

The patterns in the music can be used to define those in the dance. The identification of pattern can be a route to recording algebraically as long as the elements of the music/dance can be recorded as variables, such as classifying movements as light or strong. Here the patterns stressed could be based on the movement of feet, arms or bodies (left, right, forwards, backwards, turn or

PE: Building geometrical thinking into playing games

spin). In a similar way, repetitions of these movements need to correspond to the music. The other dance element that must be considered is speed, which is usually classified as quick or slow.

Step 1
Start the activity by watching a video of a group performing a dance routine to a song. Introduce the key ideas by identifying and recording diagrammatically the spatial arrangements that they use, noting any transitions that occur.

Step 2
The key dance elements and repetitions should also be identified, focusing, for example, on helping pupils to distinguish between light and strong movements.

Step 3
Teach pupils a notation that can be used to record their dance movements or encourage them to invent their own.

Step 4
Identify the number of people in the group and all the possible physical arrangements that could be made. Obviously smaller numbers will be easier to control and guidance may be required for lower attainers, although it will not be essential to list all possible arrangements.

Step 5
Listen to the music, identify the rhythm, and give particular attention to the chorus.

Step 6
Choose how to split the music into sections and identify the number of beats in each one.

Step 7
Identify any pauses or changes in rhythm.

Step 8
Plan the dance to fit the patterns identified in the music. The plan can initially be written out in detail using words but the patterns may be emphasised using a simple algebraic style of notation. For example, using 'L' to indicate a light movement, 'S' a strong movement and 'P' a pause, you might choose the following: 'Repeat twice (L L L S) P Repeat twice (L S L S)'

Step 9
The complete dance can then be rehearsed, performed and evaluated.

Numeracy and Mathematics Across the Primary Curriculum

Extension activities
- The activity could be carried out with music from a different time, place or culture. Explore the different movements that are stimulated by the various patterns that children identify in the music.
- Encourage pupils to explore the patterns in travelling in folk dancing or square dancing by learning dances and then finding ways of recording the movements.

Activity 3

Am I improving?

Key concepts
In designing and responding to athletic challenges pupils need to decide what makes a performance effective and then identify ways to measure changes within it. Suitable challenges and competition can motivate pupils to improve their performance by studying the experts that they admire or by looking closely at their own practice. While the monitoring of effective performance often involves studying the movement of the body in space, which may require sophisticated observational skills and kinaesthetic awareness, the outcome of athletic activity is often easily quantified by measuring distances or time.

The mathematics
The mathematics in athletics focuses on measuring distances and times. It can therefore help children to learn how to measure accurately and understand the ordering of decimal numbers. Data-handling skills can also be developed if data are collected over time and improvements in performance monitored.

The measurement of distances will arise in many situations. Pupils will have the opportunity to measure heights in centimetres for the high jump and length in tens of metres for throwing events or hundreds of metres in track events. All of these can give pupils valuable experience in ordering different types and sizes of number.

For speed and simplicity some distances can be permanently marked on the field and used as a baseline from which further measurements are taken. For example, lines marked at 10 m intervals will enable the distance that a ball has been thrown to be measured almost instantly by reference to the nearest marked line; this is quick and does not require a long tape measure (see Figure 5.1). Having essentially laid out a number line on the field, simple addition and subtraction can then be used to establish the total distance thrown.

Bar charts can be drawn to represent the performance of pupils over a number of weeks. Individual pupils can monitor their own progress or the range of performance over the whole class can be identified. The average performance could be calculated for the class and then compared with other classes or the previous year if the data have been kept. Pupils may also be interested to compare their performance with county, school and world records and can be shown this information in bar charts. It will be possible to use ratio and percentage to compare a pupil's performance with these records.

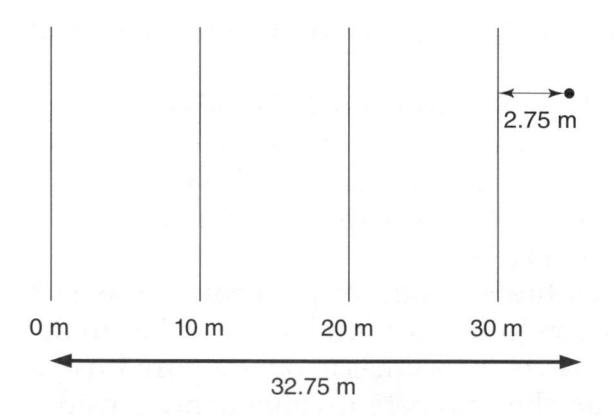

PE: Building geometrical thinking into playing games

Figure 5.1 Using markings on a field to measure the distance a ball is thrown

Step 1
Tell pupils that they are going to find out how good they are at standing jumps.

Step 2
Demonstrate to the group the way to make a standing jump.

Step 3
Working in pairs pupils monitor each other's performance and measure the distance covered to the nearest centimetre using a tape. Challenge pupils to see who can jump the furthest and/or who can make the biggest improvement during the course of the lesson or a series of lessons. The improvement could be taken as the increase in the distance jumped in centimetres or as a percentage increase.

Step 4
Pupils record the distances jumped and note how altering their jumping methods affected their performance, e.g. strong use of the arms compared to weak use of the arms. This is a good opportunity to reinforce the importance of repeating a physical task several times to find an average, rather than taking the results of one trial as conclusive.

Step 5
When you return to class, the data collected can be put into tables and bar charts. The pupils can interpret the results to see who won the set of challenges.

Extension activities
- Over the course of several PE lessons give pupils clipboards or whiteboards to make a note of their performance; they can then transfer this information into tables and graphs and monitor their performance over time. These records can be updated during registration time on the morning after the PE

lesson. The interpretation of the bar chart will have considerable personal significance for pupils!
- Working in pairs one pupil watches the other throw a ball. The observer then describes the action of the throw before modelling it to the thrower. This will stimulate the use of correct and precise vocabulary. You could ask the pupils who are observing to draw the key parts of their partners' techniques, or to emulate them using a suitable doll or 3D model.
- Running events will involve time as the measure of performance, which gives an opportunity for pupils to learn to use and read stopwatches to the nearest tenth of a second. If the stopwatch gives times to one-hundredth of a second pupils will need to round-up the numbers involved and consider why it is inappropriate to give times to one-hundredth of a second unless electronic timing systems are used.

Activity 4

The score orienteering competition

Key concepts
Orienteering is a physical activity that requires a range of problem solving, mathematical and physical skills but can be difficult to run in school as competitions require participants to complete the course one at a time. Score orienteering is a variation in which everyone participates at the same time (or at least in a large group) and the winner is decided by points scored in a fixed time rather than time taken to complete the course.

This activity is based on the key features of a score orienteering competition and requires pupils to find the best route around the school playground and/or school field using an accurate map, in an effort to score the most points in the time allowed. To complete the task pupils will need to be able to read a map successfully and use it to plan their routes around the control points. They must choose the order in which to visit them carefully and use the map to put this plan into practice. They must use their knowledge of their own stamina and speed to decide how long it will take them to complete the course so that they can finish within the time limit and avoid any time penalties. Although there is considerable overlap with geographical skills in terms of map reading and using compass points, the problem-solving context is very different. Completing the course does depend in part on the physical condition of the pupil, but unlike cross-country running there is also a significant cognitive element to this activity, which means that the fastest runner will not necessarily win.

The mathematics
Most of the problem solving in this activity makes explicit use of mathematics. Route-finding tasks enable pupils to develop their knowledge of measurement and the way ratio is used to define the scale on a map or plan as they make direct links between the distances in the playground and those represented on the map. Compass points can be used to help pupils ensure that they can orientate

PE: Building geometrical thinking into playing games

the map correctly; pupils can be assisted in this by colouring the north edge of the map blue and linking this with some feature on the northern edge of the school playground. Some aspects of planning a route can then be discussed in terms of angles or compass points. The shapes on the map must be linked to the particular features they represent, thus extending the pupils' ability to see relationships between 3D objects and 2D representations. If a scoring system is used then calculations will have to be made and timing the activity will develop pupils' ability to measure and estimate the passage of time.

This activity assumes that pupils have already had some experience of using maps or scale plans and are familiar with the general principles of orienteering.

Step 1
Set out the course placing ten distinctive markers on principal features in the school grounds and marking their position on an accurate map. You will need four or five master maps from which pupils can copy the position of the control points. Make sure you have enough copies of the unmarked map to give one to each pupil.

Step 2
Explain to pupils that their task will be to visit as many control points as possible within the time limit (e.g. 15 minutes), recording the code found at each control point.

Step 3
The pupils' first task is to copy the control points from the master map onto their maps. This is a significant mathematical activity in itself. Control points are con-

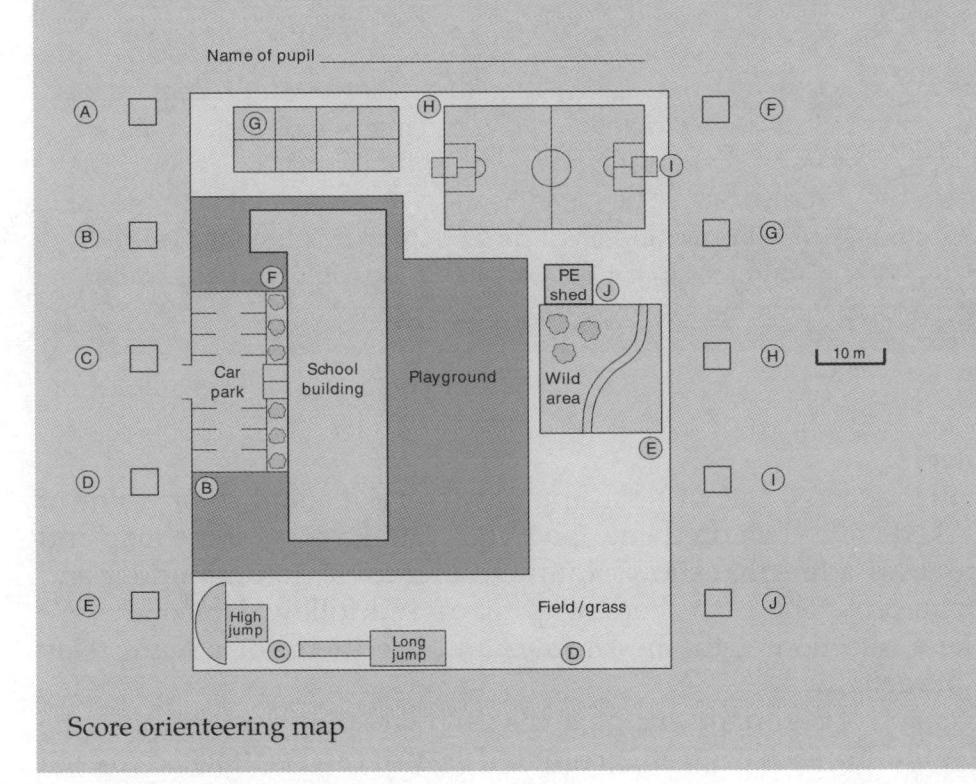

Score orienteering map

ventionally marked on the map as circles but any symbol could be used. The pupils should also label the boxes around the edge of the map A to J so that the control codes can be copied when each control is found. Each control is worth a number of points: a score of 10 per control will keep the subsequent arithmetic simple; a higher number of points for controls that are further away or more difficult to locate will make the planning more difficult. Try to arrange the challenge so that most pupils are close to achieving the maximum score.

Step 4
The pupils then need to plan their route around the control points and decide whether they will try to visit them all. At this stage identify the penalty points to be deducted for each minute they are over time; 5 or 10 points per minute will probably be suitable assuming it will take less than one minute to return from any point in the school grounds. Allow approximately 5 minutes for the pupils to make their individual plans and ask them to mark these on their maps.

Step 5
Start the whole class at the same time so that there are no significant timing difficulties. The pupils must record the code from each control point in the appropriate box on their map.

Step 6
Allow the pupils to add up their own scores. Check that all pupils have recorded the correct codes. If all pupils have achieved near maximum scores the competitive element can be minimised.

Step 7
Discuss the routes pupils chose and ask children to explain how they went about the exercise, as this will enhance problem-solving and map-reading skills and will help pupils learn how to estimate distances using scale on maps. Planning the shortest possible route between two points is clearly a key mathematical concept here, but some pupils may initially prefer to go to control points that they think will be easy to find. It is important that pupils are able to identify the correct location of the control points on the map and on the real playground.

Extension activities
- Use a plan of the school hall with the apparatus set out and the north side of the hall marked with a coloured line. Give each pupil a copy of the map and ask them to draw a line that shows an interesting route around, under and over the apparatus. They can then swap maps and follow someone else's route. Make a selection of the best routes and use them for a team relay version of the activity.
- The 'star' version of the orienteering exercise involves pupils returning to the teacher after finding each control. A coloured crayon can be attached to each

control point so that pupils, on arrival, can colour in the circles marking each control point on their map. Rather than having to follow individuals around the course, this method enables the teacher to monitor pupils' progress and to ensure that they are reading the map correctly and finding the right control point.
- If possible you can use larger areas of park or woodland to make the map reading more complex and the overall challenge greater. Start these activities by taking the pupils round a course you have set out as a group, ensuring that they can follow the map, identify the control points and mark them accurately on the map. Make extensive references to distances, scale, compass points and angles at appropriate times as you go round the course.
- Find out if there is an orienteering group in your local area who can provide resources, expertise and maps suitable for more challenging activities.

Activity 5

Gymnastics

Key concepts
The performance of gymnastic movements requires pupils to learn to distinguish elements of their movement using the key dimensions of level, speed and direction. To compose a series of movements for floor or apparatus pupils need to consider the physical skills they have and select a set of movements (including one balance) that will blend to produce a fluent sequence. They will have been taught how to classify movements according to the characteristics of level, speed and direction and will be using these features to evaluate their own and each other's performance.

The mathematics
The key mathematics in this activity is for pupils to visualise the body shape required for a particular movement and then to recognise how closely their performance matches the ideal of their visualisation. They will develop their awareness of the 3D space in which the movement takes place and make extensive and accurate use of the vocabulary of shape and space to describe each other's movements when they give feedback. The key concepts involved are symmetry and asymmetry, ways of describing rotation, and the use of 3D space including the relationship of the moving body to the vertical and horizontal. There will be opportunities to use much vocabulary in context although the strict mathematical meaning may be stretched a little: triangle, square, vertical, horizontal, straight, pointed, curved, rounded, right angle, full turn, half turn and measurements in degrees are just a few examples.

Numeracy and Mathematics Across the Primary Curriculum

Step 1
The pupils get together in pairs for this activity so that they can give each other feedback on their performance; most pupils will be unable to recognise their own body shape during movement.

Step 2
Set the task clearly and identify a number of examples of balances and other movements that the pupils have previously been taught. Emphasise the quality of movement that is sought by giving examples of good movement using descriptive language, photographs, diagrams or video.

Step 3
Pupils then work in pairs to prepare their sequences, first trying out the moves and then receiving feedback from their partner in verbal or sketched form.

Step 4
Ask the pupils to record their sequences as a set of diagrams and to identify key features of each stage that represent examples of quality movement. They can also write a description of the movement sequence to identify what makes the performance effective.

Step 5
The pupils rehearse the movements and try to make improvements while continuing to receive feedback from their partners. If pupils are willing, it would be worthwhile for them to perform their sequences to the class and receive feedback from a larger audience.

Extension activities
- Prepare a display using the diagrams for the best sequences along with a selection of evaluative comments that identify what was particularly effective about the sequence.
- Use digital cameras or video to record pupils' movements so that they can have this visual feedback on their performance. This can then be compared with the language that is used to describe the movements and the shapes achieved.

6 Music
Notes and number

Music is a powerful and unique form of communication that brings together human thought, creativity and action, feeling and sensitivity. Today, music is as much a part of any culture as it was in the past; it helps us explore attitudes and values that inform societies, understand cultural change and develop physical and perceptual skills. As a subject it is so much more than a series of concepts, rules and theories that are not open to personal interpretation. Music, like art, is largely subjective and stimulates personal expression, reflection and emotional development. Listeners' opinions on musical quality are influenced by personal taste and opinion rather than objective reasoning. It is not just a collection of organised sounds, although certain key elements can be identified: pitch, duration, dynamics, tempo, timbre, texture and silence combine in various ways to create different musical structures.

Where is the mathematics in music?

Like mathematics, much of the structure of music is based on patterns and relationships. The patterns, relationships and sequences of sounds, notes and rhythms in music resemble those of numbers, shapes and measurements in mathematics. The two disciplines are even more directly related since the duration of musical notes and the intervals between them are measured and defined in terms of numerical values. For some, mathematics represents the essence of music making and composing.

The point at which mathematics and music clearly merge is where audible sounds are characterised in abstract concepts such as number. Mathematics can highlight relationships between sounds and rhythms, which can then be turned into audible patterns and structures of music. Children can use mathematical processes to arrange sounds in different orders and patterns. Just as mathematical relationships can be shown using tactile objects, such as Unifix or Dienes apparatus, so too can musical relationships be demonstrated using sounds.

Music is probably the closest subject area of the National Curriculum to mathematics, but the links between the two are not always clear and need to be made explicit for the benefit of both areas of experience.

Listening and applying knowledge and understanding

Within performing, composing and appraising, children need to be taught how musical elements can be organised to communicate different moods and effects. These musical elements and their accompanying mathematical concepts are shown in Table 6.1.

Table 6.1 Musical elements and their corresponding mathematical concepts

Musical element	Mathematical concept
Pitch: high and low sounds moving upwards and downwards by steps and leaps	Pattern, relationship, measurement of height, number, interval, proportion, coordinates, average, frequency
Duration: rhythm patterns (e.g. beats grouped in twos, threes and fours)	measurement of time, frequency
Dynamics: different levels of volume (e.g. loud and quiet; getting gradually louder/quieter)	pattern, informal measurement, proportion
Tempo: different speeds (e.g. fast and slow; getting gradually faster/slower)	measurement of speed, number, frequency, fractions, proportion
Timbre: how one sound differs from another	attributes, ordering, number
Texture: building up rhythm patterns, adding layers of sound (e.g. adding accompaniments to songs)	pattern, interval, repeating patterns, number
Structure: the overall shape/form of music	shape (e.g. circles), symmetry, asymmetry, repetition, curves, other transformations, representing data, probability, frequency

Activity 1

Creating and developing musical ideas: Can we compose a piece of music that contains a rhythmic reflection?

Key concepts
A number of musical concepts underlie this activity. Composition is a type of communication – it helps people to express their feelings. The 'vocabulary' in this activity is made up of pace, duration, structure and timbre.

The mathematics
The main mathematics concept in this activity is symmetry of sound. Reflection is defined as a mirror image and a transformation; a rhythmic reflection is one where the sound is symmetrical – the same beat is returned as a mirror image.

Step 1
With children working together in pairs, begin by asking them to use body percussion (e.g. slapping hands on thighs, finger-tapping on heads, rocking feet heel-to-toe) to create simple rhythms for their partners to copy. (Clapping doesn't supply as many visual clues as other forms of body percussion and could be used at a later stage. Some children will need lots of practice to be able to copy their partners' clapping patterns.)

Step 2
When the children can confidently copy each other's patterns they can then respond with a reflection.

Step 3
The children then compose a piece of music using an identifiable reflection as a feature.

Extension activities
- This can also be played as a circle game with one child creating a pattern and the others responding one at a time with reflections.
- A further stage to attempt before composing is to do the same activity with pitched instruments.

Activity 2

Food music: Compose and write down a composition without using standard notation

Key concepts
In this activity children create a composition using the rhythmic patterns made by words relating to the theme of 'food'. Obviously the choice of topics and the words relating to them are limitless – food is suggested here because the children will be able to think of lots of examples. Notation is the way music is written down. The form that this takes can vary from standard musical notation to non-standard notation, which employs graphic means of conveying the composer's intention to players. In this activity words are used as notation as they are the vehicle for pace, duration structure and pitch.

The mathematics
Rhythms lend themselves easily to work on fractions and the associated concept of division. Whole 'units' (bars) of music are broken up into notes or non-standard notation that have a value in terms of beats. Repeating patterns of rhythms also correspond in many ways to algebraic concepts. A matrix is used in this activity to represent patterns of rhythm.

Numeracy and Mathematics Across the Primary Curriculum

> *Step 1*
> The children work in groups and devise a rhythmic composition, recording their notation (using words or pictures) on a four-by-four grid.
>
> *Step 2*
> It's easiest if the children work on their rhythms within a 1-2-3-4 pulse or beat. Getting them to count these beats aloud at the start can be helpful.
>
> | | 1 | 2 | 3 | 4 |
> |---|---|---|---|---|
> | 1 | Green | apple | green | apple |
> | 2 | birthday | cake | cheese | — |
> | 3 | Baked | beans | buttered | toast |
> | 4 | nice | mug of | tea | — |
>
> Simple grid notation

Extension activities
- These compositions can very easily be turned into raps that will entertain the children immensely!
- If the children are ready they can begin to add pitch using a limited number of notes (e.g. A G E). They will be excited to learn that by adding pitch they are going through the process of writing a song.
- Once the children have gained confidence, introduce the standard notation as appropriate. The advantage with this activity is that, musically, children are getting to grips with the idea of breaking music up into bars before they work with standard notation. Mathematically they will be thinking about fractions.

Activity 3

Notating polyrhythms using a matrix

Key concepts
This activity is related to Activity 2 in that initially the children will find a simple four-by-four grid and a 1-2-3-4 pulse easier to work with. This activity explores the idea of notating rhythms using simple symbols. Polyrhythms are a type of composition that need considerable concentration and sense of

Music: Notes and number

rhythm. Highly intricate compositions and harmonies can be produced from the underlying mathematical patterns that determine them.

The mathematics
Here the main mathematical concept is repeating patterns of groups of numbers, or in musical terminology, 'counted sections'. So, '1-2-3' is a 'counted section of 3'. Prediction will also be an important skill that children will need to employ when the patterns become complex. Time signatures will help introduce the idea of fractions: a 'counted section of three' works as three thirds in terms of beats.

Step 1
In this activity the whole group will be using a 'counted section of four' with each beat being 'one quarter' at the start of the activity. The group begins by counting '1-2-3-4' and clapping each beat, preferably accompanied by a drum to give a steady pulse.

Step 2
Still using '1-2-3-4' get the group to clap the first beat, rest (stay silent) on the second and divide the third in two to make eighths.

Step 3
Develop with the second beat being halved.

Step 4
Halve all the beats.

Step 5
Choose suitable symbols that the pupils can use to notate this rhythm (e.g. X = whole beat, x = half beat, 0 = rest). The example below is very simple; the more experienced children can make more complex grids.

	1	2	3	4
1	X X X X	X X X X	X X X X	X X X X
2	X O xx X	X O xx X	X O xx X	X O xx X
3	X xx X X	X xx X X	X xx X X	X xx X X
4	xx xx xx xx	xx xx xx xx	xx xx xx xx	xx xx xx xx

Notating polyrhythms using simple symbols

Extension activities
- Use a 'counted section of three'. Clap on the first beat and rest on beats two and three, then repeat. Use a 'counted section of four', again clapping on the first beat, resting on the remaining beats and then repeating the section. Divide the group in half and beat on a drum to provide a steady pulse. Give one half of the group 'three' and the other 'four' and start them off at the same time. This will result in claps sounding together, sometimes consecutively and sometimes not. Mathematical prediction can be used to establish when both halves will clap together: $3 \times 4 = 12$. This pattern can be explored as the activity continues in duration. Other sequences can be predicted and explored (e.g. $3 \times 4 \times 5$).
- For variation, children can begin to include accents and time signatures, which will again involve fractions: the top figures represent the number of beats to each bar and the bottom, the length of the beat. These will need to be kept very simple to ensure that the children do not get confused:

   $\frac{2}{2}$ two minim beats in a bar (i.e. two halves of a semibreve)
   $\frac{2}{4}$ two crotchet beats in a bar (i.e. two quarters of a semibreve)
   $\frac{3}{4}$ three crotchet beats in a bar (i.e. three quarters of a semibreve)
   $\frac{4}{4}$ four crotchet beats in a bar (i.e. four quarters of a semibreve)

- A useful way of demonstrating time signatures is to relate them to dances, e.g. Waltzes, polkas, reels, jigs and so on.
- Investigate the number of rhythms that can be made using different combinations of notes (e.g. crotchets and quavers, or crotchets, quavers and minims). Ask the children to look for patterns in their compositions. As they become more experienced, new kinds of notes can be introduced (e.g. semibreves and semiquavers) and rests can be used to indicate intervals of silence in their musical compositions.

Activity 4

Keeping time: Can we make a pendulum that gives exactly 60 beats per minute?

Key concepts
This activity links three subject areas together:

Science: Children will be exploring various scientific concepts including the harmonic, continuous motion of the pendulum as the energy is used and it comes to a standstill. The length of a swing (beat) will depend on the position of the weight on the string. The children should be encouraged to experiment with different beats and then record their findings on a graph.

Maths: Time is important in music not simply as far as tempo is concerned, but also to divide musical pieces into equal sections or to measure the length of time

Music: Notes and number

that voices or instruments can sustain a note. The vocabulary used to compare these sounds can start off quite generally (e.g. with simple references to sounds being 'longer' or 'shorter'). As children begin to grasp the principles of measuring, they can actually time the length of notes using formal measures such as counting (using seconds and minutes) and informal measures such as clapping, counting number of breaths and so on.

Music: Recording environmental sounds on a tape recorder will offer the children the opportunity to make parallel timelines of the sounds. This allows linear measurement to be related to the passage of time.

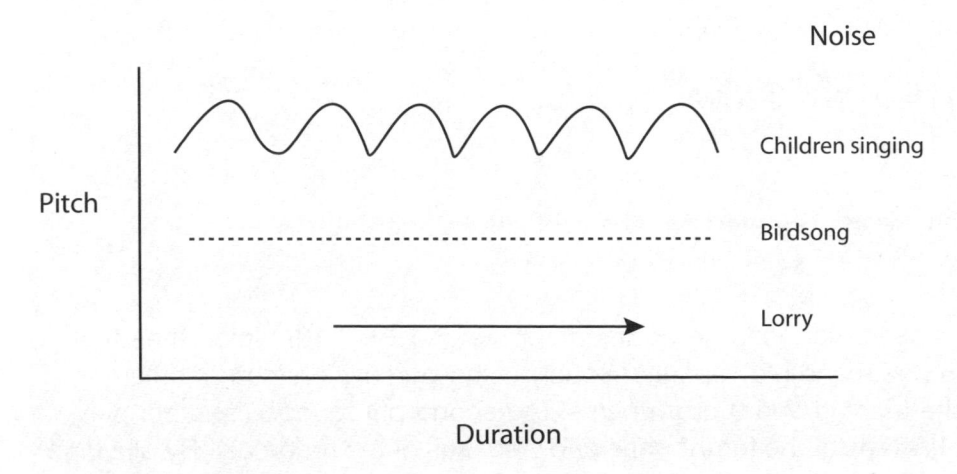

Figure 6.1 Timelines of environmental sounds

The children can transform the tape-recording into a musical piece, choosing instruments that best represent each of the sounds and timing their entries and the length of time they play for accurately. As a prelude to their performance, the dynamics and pitch of the sounds can also be illustrated using graph paper.

Activity 5

Using the patterns in number to compose music

Key concepts
Music is largely about repeated patterns that include the basic elements of pitch, duration, dynamics, tempo, timbre, texture and silence. This activity uses mathematical patterns as a basis for composition and performance.

The mathematics
Number bonds and progressions have excellent rhythmic qualities. Here the addition and subtraction facts up to ten are used.

Numeracy and Mathematics Across the Primary Curriculum

Step 1
Addition and subtraction facts up to ten (or beyond) are charted on a grid.

```
            1  2  3  4  5  6  7  8  9  10
1 + 9       ○  □  □  □  □  □  □  □  □  □

2 + 8       ○  ○  □  □  □  □  □  □  □  □

3 + 7       ○  ○  ○  □  □  □  □  □  □  □

4 + 6       ○  ○  ○  ○  □  □  □  □  □  □

5 + 5       ○  ○  ○  ○  ○  □  □  □  □  □
```

Examples of addition facts up to ten

Step 2
Compositions are then created and the grid used as notation.

Step 3
In pairs, the children choose an addition sum (e.g. 1 + 9 = 10) which they then translate into beats and play simultaneously using percussion instruments (e.g. 1 strike on the triangle and 9 drum beats). A second pair choose another sum, say 3 + 7 (e.g. 3 taps of the tambourine and 7 shakes of the maracas). By playing these lines simultaneously an interesting texture will be produced.

Composition based on addition facts up to ten

Step 4
The composition will last for just ten beats but for added interest, the concept of a round – a circular shape – can be introduced through repetition. The first pair of children begin at Line 1 and as soon as they have reached Line 2 the second pair start at Line 1. Carry on until each pair has played the whole grid. The texture created will be interesting – it will get richer as more children join in, and after swelling in the middle the sound will die down and eventually leave a single pair playing a 'solo' at the end. You can illustrate the concept of the round diagrammatically, and even start to introduce fractions.

Extension activities
- The order of the lines can be changed to see if a more pleasing effect can be generated:
  e.g.

	1	2	3	4	5	6	7	8	9	10
2 + 8	○	○	□	□	□	□	□	□	□	□
8 + 2	□	□	□	□	□	□	□	□	○	○

- Use the beats to start to introduce fractions: $8/10$ths are drum beats; $2/10$ths are tambourine taps.
- Try subtraction using the number bonds with two instruments starting together, one playing ten beats the other playing two. After two beats one of the instruments will stop and the remainder of the beats by the other instrument will denote the difference between 10 and 2, i.e. 8

Activity 6

Rhythmic patterns

Key concepts
This activity focuses children's attention on repeated numerical patterns and notation. It encourages children to develop the skills needed to play different rhythms together.

The mathematics
Multiplication tables are extraordinarily effective sources for rhythmic patterns. This activity is particularly effective in introducing the link between rhythm and factor and multiples.

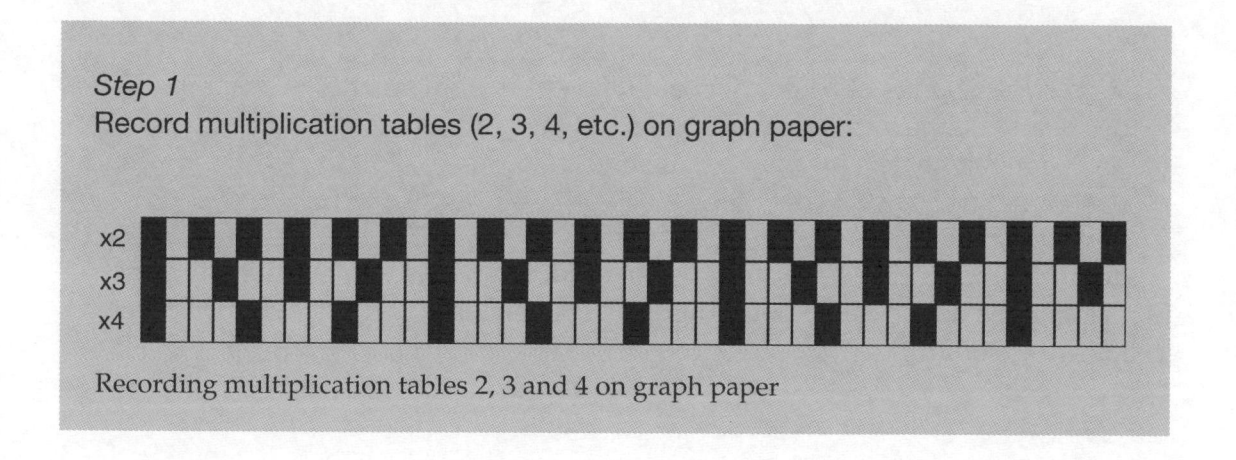

Step 1
Record multiplication tables (2, 3, 4, etc.) on graph paper:

Recording multiplication tables 2, 3 and 4 on graph paper

Numeracy and Mathematics Across the Primary Curriculum

> *Step 2*
> Give each child an unpitched instrument. Allocate a separate line of the multiplication table to each child and ask them to play when the counted pulse reaches a filled-in box.
>
> *Step 3*
> Let a number of children choose different consecutive tables and then play them together. They might predict when each of the instruments would sound simultaneously and use the chart to 'prove' this.

Extension activities
- Repeat the exercise but this time replace the multiples of 2, 3, 4, etc. with prime numbers or factors. What happens if the children play the patterns from 3, 6, 9 times tables? What happens when they play squared and cubed numbers?
- Vary the rhythm by playing to a given speed, then twice as slow then twice as fast.

Activity 7

Performance: Controlling sounds through singing and playing

Key concepts
In this activity children will focus on developing a firm control of sounds using very basic notation.

The mathematics
This activity combines many of the mathematical skills and concepts that have been discussed earlier on in this chapter: counting in 2s, 3s and 4s, recognising patterns, recording and repeating them.

> *Step 1*
> Begin by asking the children to create different rhythmic patterns using three or four beats. Suggest that they use the word 'long' to represent whole beats and 'short' for half beats.
>
> *Step 2*
> Produce simple notation to show this and get the pupils to make sure that each rhythm adds up to three or four. This is also an excellent way of introducing fractions to children.

Music: Notes and number

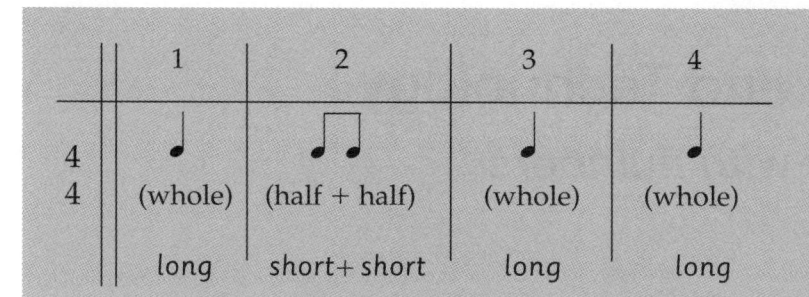

	1	2	3	4
4/4	♩	♫	♩	♩
	(whole)	(half + half)	(whole)	(whole)
	long	short + short	long	long

Simple notation using four beats

Step 3
Introduce the idea of intervals of silence in the music (rests). A rest is very much like a 'zero' in mathematics in as much as it indicates a silent beat or 'no sound'.

Step 4
Use percussion instruments, such as drums, blocks or tambourines, voices or body percussion, to develop this activity further by varying the rhythm.

Extension activities
- Write the notation on flash cards so that different sequences can be made and rearranged easily. This visual stimulus will help children to spot patterns too. Once some patterns have been created and recorded using simple notation (i.e. words or pictures) they can be linked together to create even more sequences.
- See if the children can create patterns that are symmetrical:

short – short – long – long – short ‖ short – long – long – short – short

- As children gain experience of creating and recognising patterns there should be opportunities to work on prediction skills – guessing which patterns will repeat themselves.

7 Design and Technology
Creating with numbers

Unlike most other subjects in the National Curriculum, design and technology (D&T) cannot be classed as a single subject area. Pupils are required to apply a mixture of scientific knowledge and understanding, physical aptitude and problem-solving skills to design and build products or systems that people want. In this sense D&T is more than just applied science.

At the start of a D&T project pupils must identify a need, want or opportunity for a product or system in their local environment. Once this 'gap' has been found pupils can launch into the 'design' aspect of the project. Here they must combine their creative response to the problem with concerns for functionality, aesthetics, the environment and ergonomics. The implementation of all of these ideas forms the 'technology' aspect of the project. Pupils are expected to use their scientific knowledge to construct products or systems that will function effectively.

In summary, the D&T curriculum identifies a five-stage problem-solving process, which involves:

- Identifying the need
- Generating design ideas
- Planning to implement the design
- Implementing or constructing the design
- Evaluating the product or system

Children are expected to encounter D&T activities in a broad variety of contexts, covering electrical and mechanical components, food, mouldable materials, stiff and flexible materials and textiles (DfEE/QCA 1999). The activities tackled within these contexts should give children opportunities to investigate and evaluate familiar products (possibly by taking them apart), to focus on practical tasks that involve 'skill-getting' (i.e. developing a range of techniques, skills, processes and knowledge), and to design and make products or systems that people want.

Knowledge base of the subject

The problem solving nature of D&T is given a high profile in the National Curriculum (2000) for England where it is suggested that the subject 'calls for pupils to become autonomous and creative problem solvers' (DfEE/QCA 1999: 90). The National Curriculum identifies this aspect of D&T as 'developing, planning and communicating ideas' and 'evaluating processes and products' (p. 92). These elements have considerable overlap with the model of problem solving that we have used in this book (see Figure 3 in the Introduction). Mathematics can make a useful contribution to the understanding of D&T at any stage in the process.

The development of the necessary skills to use tools effectively is also part of D&T. This covers the use of basic tools for working with food, paper, card, wood and plastic, and the decision making behind choosing appropriate methods or suggesting alternatives if an approach is unsuccessful.

Scientific understanding is clearly a significant element of the knowledge used in D&T, and technology may be viewed as the application of scientific ideas to solve problems. The materials aspect of science will be particularly significant for evaluating and making products. While exploring the qualities of materials the children will examine how well they fit their purpose/function and how their properties can alter as the materials are used to make particular products. The physical processes element of science will be relevant in the use of electricity in the design of products. Forces and motion will be of interest when children consider how products need to be made to withstand the forces they may be subject to in use, or to explore the way in which products fail. The properties of light and sound may be key attributes of some designs.

The pupils' understanding of these aspects of science will, to some extent at least, guide their understanding of the technical possibilities available when they undertake their own design and technology assignments. Some aspects of life processes and environmental issues will be key features in the evaluation of products' environmental impact. All of these scientific topics may in themselves involve a significant element of mathematical thinking in developing the pupils' understanding.

Breadth of study

The design and technology activities that you undertake with your pupils are expected to cover a variety of environments including the home, school, community, recreation, business and industry. Within these environments the National Curriculum suggests that pupils tackle tasks that involve:

- investigating and evaluating familiar products;
- practical elements designed to develop particular skills, processes or knowledge;
- assignments that enhance designing and making skills.

Where is the mathematics in D&T?

When pupils are using tools to build their designs or carry out a focused practical task they will make extensive use of their measuring skills and will work with a variety of instruments. The need for a particular degree of accuracy and the approximate nature of all measurements will figure significantly in their work. Exploring the use of pulleys, levers, cogs and gear ratios will cover many of the basic aspects of the mathematics (and science) of mechanics. Much number work can be done on the concepts of ratio, factors and multiples as pupils try to work out the effects of different sets of gears on the speed of an electric motor, or the effect of different arrangements of levers and pulleys. The use of switches and sensors in control technology will lead to the planning of automated sequences of events, which could include the mathematical use of flow charts and networks.

Pupils' knowledge of basic geometry will come into play when they analyse structures in which the properties of triangles, right angles and rectangles can be linked to rigidity and strength. As pupils try to communicate their design ideas on paper they will see that 3D objects can be represented in a variety of 2D formats. Isometric and scale drawing (plan and elevation) may be used as pupils develop design ideas and try to explain and communicate them clearly. Putting a design onto paper may also be a good opportunity to enhance pupils' understanding of the level of accuracy required when measuring in different circumstances, and to explore the use of ratio in scale drawing. Identifying the sequence of actions needed to build a design and suggesting alternatives will lead to creative problem solving in which the use of flow charts and networks is helpful. When pupils carry out tests to evaluate the quality of a product or process they will need to employ a combination of measuring and data handling.

Activity 1

Design and build a crane

Key concepts
We begin with a focused practical task to build a crane using a wooden structure, an electric motor and gears or pulleys. The children will have the chance to develop skills in designing a rigid structure using wood and the techniques required for its construction. Their understanding of how things can be made to move using a geared mechanism and the use of electrical circuits with simple switches will also be enhanced. Giving pupils a complete design to build, thus focusing attention on the construction aspects of the task, can provide support and make the task easier. Alternatively pupils can be challenged to design some parts themselves. To make the task more challenging, targets can be set that will be used to evaluate the products once they have been built. These might include the height the crane must raise a particular weight, the length of the arm of the crane, the accuracy with which a load can be placed, or the aesthetic qualities of the design.

The mathematics
To design the crane structure the children will need to apply their knowledge of the properties of 2D shapes to solve problems. They will also need to employ their skills in using 2D drawings to represent 3D shapes and their ability to draw accurately. In particular they will need to appreciate the use of the triangle to make a rigid structure. A geared mechanism will be needed to reduce the electric motor's speed of rotation so that the load is raised at an appropriate speed and a higher maximum weight can be lifted. The design and development of this system offers opportunities for introducing calculations that incorporate ratios, factors and multiples, either exactly or using approximations. Where pulleys are used the measuring or calculation of circumference will arise. The need to balance the load lifted with counterweights will introduce the principle of levers and may lead to additional calculations, again either exact or approximate.

Step 1
Start by explaining the task and the main design criteria to the pupils: 'Design a crane to raise a 50 g weight through a height of 30 cm.' Provide the children with a prototype for the gearing mechanism, motor and switch, and explain how it might be modified. This can use gears or pulleys (cylinders) depending on the resources available.

Step 2
The pupils should start by exploring the construction of rigid structures using straws and pipe cleaners. They should also study pictures of cranes to identify the key features of the design. Make sure that they identify the counterweight and help them to understand its purpose by demonstrating the principle using a model that over balances.

Step 3
Pupils should then make drawings using a combination of two-dimensional plans and isometric representations of the three-dimensional objects they are designing. At this stage the constraints on the materials they can use need to be explored. For example, they might be limited in the amount of wood they can use overall, or the maximum length of one piece of wood might be set at 20 cm if the task were to design a crane to lift a load a maximum distance of 30 cm.

Step 4
Pupils check and revise their designs if necessary and complete plans for installing the gear mechanism, showing the position of the electric motor and switch.

Step 5
They then make the structures they have designed. Encourage them to measure accurately to ensure that their structures are vertical.

Numeracy and Mathematics Across the Primary Curriculum

> *Step 6*
> Pupils add the gear system, electric motor and lifting mechanism. The speed of the electric motor will need to be reduced to provide an appropriate lifting speed and to lift greater loads; this can be done by trialling and gradually improving methods or by calculating ratios to make the process more controlled. If they are using commercially available sets of gears, their ratio calculations could be exact. If on the other hand they are using different diameter cylinders to make pulleys, their calculations will be approximate.
>
> *Step 7*
> The mechanism is tested to see that the crane runs at an appropriate speed and to find the maximum weight that can be lifted. If adjustments are needed, start to talk about ratios to encourage the pupils to calculate how much faster or slower these will make the crane: 'If you double the circumference of the final pulley in the drive, the load will move at half the speed.'
>
> *Step 8*
> The final product(s) can be evaluated against the set criteria – possibly including an aesthetic element – and the key areas for further development identified. This could involve the use of a preference matrix and rating scales if a variety of designs are to be compared.

Extension activities
- Use materials other than wood (e.g. paper, art straws) and do not use an electric motor. The pupils can design a winding mechanism that they can turn by hand. Ask the pupils to calculate how many turns of the handle will be required to raise the load through 30 cm and challenge them to adjust the gearing so that the lift is completed in three to four turns. The targets for the height and weight to be lifted must be modified to suit the building materials used. If several models have been built, the relationship between gearing and the maximum weight that can be lifted might be explored and graphs of the results drawn to identify the relationship.
- Design the gearing mechanism so that different gear ratios can be set up easily and the operators will be able to choose either greater speed in lifting or higher maximum weight. This can be achieved by having different circumference pulleys on the same axle or by using a construction kit with plastic gears that can be changed easily. This might be part of a competition to complete a number of tasks in a set time. Designing something less complex, like a drawbridge for a model castle, can also involve work on gearing or pulley systems as well as some basic analysis of rigid structures. Challenge children to make a lift and to design mechanisms that stop it accurately at each floor. Sliding doors could also be incorporated into the design. Both of these aspects will require accurate measurement.
- Help pupils to design and make a simple motorised four-wheeled buggy.

This will encourage pupils to consider the relationships between the size of gears or pulleys in the drive mechanism and the final speed or power of the model. Once again the mathematical focus here is on ratios. By setting a challenge to see who can operate the buggy at the fastest speed over a 5 metre course, the pupils will have the opportunity to work on the accurate measurement of time and the ordering of numbers with decimal places.

Activity 2

Which structure is strongest?

Key concepts
This activity covers a number of focused practical tasks relating to strengths in structures, particularly bridges. The children will have the chance to develop skills in designing a rigid structure using card and the techniques required for its construction. Their understanding of the forces of compression and tension will also be enhanced. The concept of strength-to-weight ratio is introduced in considering the properties of different materials and structures.

The mathematics
The mathematics involved here will cover measuring of length and weight and calculation involving division, fractions and ratios. Careful measurement will be needed to shape the materials so that they all have the same weight. This will help pupils develop their estimation skills for different materials. Having completed the experiments they can calculate the strength-to-weight ratio using division. The numbers involved will result in answers that incorporate fractions or decimals. The difficulty of the division calculation will depend on the weight of the materials that are being used: a weight of 10 g (or 100 g), for example, would produce the simplest calculations. It may be appropriate to use calculators to complete more complex division calculations and concentrate pupils' attention on interpreting the results – putting the results in order from smallest to largest and considering the differences of the strength-to-weight ratios calculated. This will help children to develop their understanding of the relative size of numbers as they extend their concept of number to include real numbers. They could also practise rounding the results of calculations to one or two decimal places. Draw attention to the approximate nature of measurement by encouraging children to think about the changes in results that might be produced if weights were added to the bridge in 1 g increments.

Step 1
Give the children a range of materials from which they might make their bridge, e.g. thin card, thick card, polystyrene sheet, balsa wood sheet, thin plastic sheet or thin plywood.

Numeracy and Mathematics Across the Primary Curriculum

Step 2
Ask pupils to find the starting weight of the material. They should use an appropriate tool to trim the material so that it weighs approximately 15 g and has a length of about 14 cm.

Step 3
The pupils then set up two wooden blocks 10 cm apart and place one of the materials to be tested across them to make a bridge.

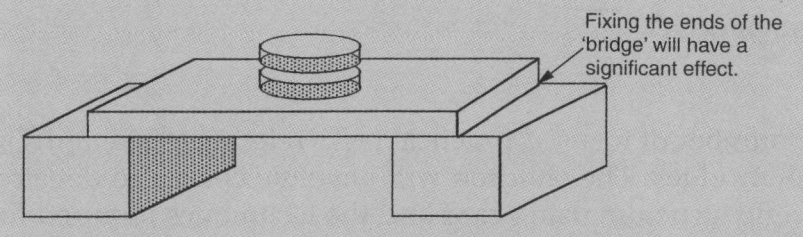

Fixing the ends of the 'bridge' will have a significant effect.

Testing weights held by different materials

Step 4
They then load 5 g weights onto the centre of the bridge until it collapses, recording the weight it succeeded in supporting in a table of results (see below). Collapse might be defined as a deflection of more than 1 cm at the centre.

Step 5
This procedure is repeated for all the materials being tested.

Step 6
The strength-to-weight ratio is then calculated by dividing the weight carried by 15 (the weight of the material). Which is the strongest material?

Material	Weight of Materials (WM)	Weight of Load carried before collapse (WL)	Strength-to-weight ratio (WL ÷ WM)
Thin card	15 g	30 g	30 ÷ 15 = 2
Thick card	15g	45 g	45 ÷ 15 = 3

Table of results

Extension activities
- Glue reinforcing strips on the underside of the bridge to make paper or thin card bridges stronger: try simple L-shapes, triangles and more complex interlocking structures. Get the children to test them in the same way as in the main activity in order to calculate their strength-to-weight ratio. As it will be impossible to use the same weight of material in each construction it will be necessary to weigh each completed structure accurately.

D&T: Creating with numbers

- The children use a sheet of thick card (say 20 cm by 2 cm), first fixed horizontally and then fixed vertically. They suspend the 5 g weights using a piece of thread so that the strength can be tested (see Figure 7.1). As the maximum weight that can be held is exceeded the beam will buckle on the top edge. Invite children to explore ways of strengthening this edge by adding further strips of card to make a 'T' beam or 'I' beam, a box girder or alternatively using a tube.

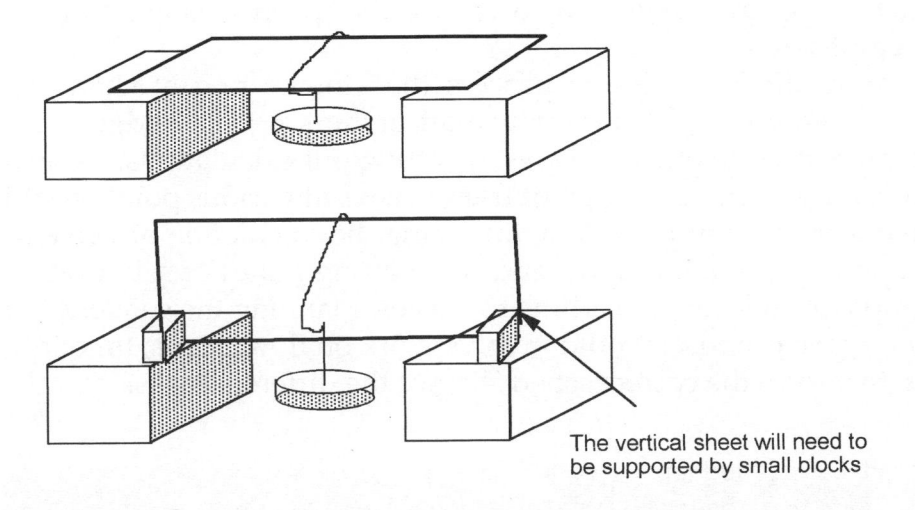

The vertical sheet will need to be supported by small blocks

Figure 7.1 Testing the strength of thick card

Activity 3

Developing the school playground

Key concepts
This activity focuses on developing, planning and communicating ideas as the pupils explore ways of improving an aspect of their local environment. As pupils think about the school playground they need to be encouraged to consider all the people who use the area, possibly including adults as well as children. As their ideas develop pupils must be encouraged to communicate them clearly using a variety of media. The activity may be limited inasmuch as alterations specified in their designs might require the use of tools and materials that are beyond their capabilities. If this is the case the evaluation of the finished work could be given greater significance.

The mathematics
The main use of mathematics is likely to be in the area of shape, space and measures as pupils try to communicate the images they have in their heads. The use of 2D to represent 3D objects can be tackled with diagrams that make use of the concept of isometric projection or through scale drawings. The activity will

involve accurate measuring of the playground using metre sticks, tape measures or tools like a trundle wheel. The conversion of the measurements to a scale will require the pupils to use calculation skills. If the conversion is kept simple by giving pupils squared paper and asking them to use one square to represent one metre, the children should still be made aware of the calculation involved and the idea of reducing the measurements by dividing by 100. This is also a good opportunity for them to grasp the expression of the scale as a ratio of 1:100. The idea of using the scale drawings and scale models of the playground to explore possible improvements and to ensure that proposals are suitable should also be emphasised.

If cost constraints affect the design process then the costing of changes/improvements will require significant mathematical thinking. If a budget is set pupils may be introduced to efficient ways of setting out calculations, using a spreadsheet for example. The collection of users' views at various points could lead to extensive work on handling data, involving the calculation of percentages, drawing and interpreting graphs, and considering issues of selecting an appropriate sample of children and adults to survey. Planning the sequence of work to implement the proposals could involve work on flow charts, timelines or Gantt charts, to ensure the correct scheduling of the different tasks.

Step 1
As a group, identify the possibility of changing the playground and set out the constraints that will need to be considered (e.g. funds available, timescale, etc.).

Step 2
The children brainstorm to find out what needs or wants should be considered (e.g. quiet area, ball games area, etc.).

Step 3
They then measure the playground and make scale drawings or models of relevant areas. Photographs and ICT might usefully be exploited here.

Step 4
A survey of the rest of the school is undertaken to find out how they view the group's identification of needs and wants.

Step 5
The children can also ask the adults in the school for their views on problems in the playground. Those asked might include the head teacher, teaching staff, dinner ladies and parents. Analyse this data together and select a need that can be met within the constraints.

Step 6
The group should then identify the design criteria that need to be fulfilled.

Step 7
Hold a design competition to produce a variety of plans to meet the need. These plans might be pictures or diagrams developed from the scale drawings already produced, or 3D models constructed from modelling clay, card or reclaimed materials.

Step 8
Judge the designs against the criteria. If appropriate make further scale drawings or models to test out the best designs and approve the winning design with the rest of the school, the school governors or the PTA if their funds are being used.

Step 9
Work with the pupils to plan the sequence of work for implementing the winning design, identifying the order in which the tasks must be completed and allowing appropriate time for each stage using flow charts and timelines.

Step 10
Implement the design involving the pupils in the measuring and construction as much as possible.

Step 11
Evaluate the finished product against the original criteria and by surveying the pupils and adults who use the playground.

Extension activities
- Run a similar project for the development of an area of the school grounds, for example a garden for the autumn, a wildlife area, or a pond.
- A smaller project could involve reorganising the pupils' classroom or developing a specific area of it, for example the computer area, book trays, store cupboard, or the cloakroom.

Activity 4

Masks for a play

Key concepts
School plays or even class assemblies often require some simple props, which provides a context for this activity. A variety of methods can be used to construct masks and you can adapt the activity to use alternative construction techniques. The main focus is on the construction phase in which pupils learn to use tools for working with paper and card as they construct their design for an animal head. They will be measuring, marking out and cutting paper and card, using glue and staples to assemble and join the components. They will use dif-

ferent coloured paint, wax crayons or felt tip pens to finish the masks and to create textures appropriate to the animal they have chosen. As they work, the pupils should be encouraged to consider the characteristics of the paper and card, identifying what makes the structure stiffer and how the materials can be made to curve. The effect of finishing the masks with different materials can be explored by comparing different finished masks.

The mathematics
The activity focuses pupils' attention on the need for accurate measuring using standard or non-standard units. It will also stimulate pupils to visualise 2D and 3D shapes as they construct the basic mask and later add the detail (i.e. the features) to the animal face. Their knowledge of the properties of shape will be tested as they try to visualise the 3D version of the mask and work out how to place the eyes, nose and mouth accurately. Some use of symmetry may arise as they consider the features of the mask and evaluate the finished product.

Step 1
Pupils should work in groups of between two and four. Each group is given a commercially produced basic mask to examine. By trying this on they can establish the key measurements that need to be taken to produce a well-fitting mask. These include the gap between the eyes, length and width of nose, width of face, length of nose, distance from chin to top of forehead, width of mouth, distance of mouth below eye line, and distance around face from ear to ear.

Step 2
Working in pairs they should prepare a sketch and then measure and record these key measurements, either in centimetres or using a non-standard measure like the width of their eye.

Step 3
These measurements can then be marked onto a sheet of paper to construct a prototype. Accuracy in measuring and relative alignment (this could be done by measuring angles but will probably be done by 'eye' only) will be very important to achieve a good fit. Holes need to be cut carefully for the eyes, mouth and below the nose.

Step 4
Identify the points on each side of the mask where it will be fixed to the head, probably using elastic behind the ears or around the back of the head.

Step 5
When the appropriate adjustments have been completed a version can be made on thin card. Encourage the pupils to be accurate in their measuring and cutting. Add the chosen fixing.

Step 6
The basic structure can then be embellished by adding detail – made in thin card or paper – to the eyebrows, nose and hair to make the mask more realistic or fantastic.

Step 7
The final stage is to finish the mask with appropriate colours.

Extension activities
- For a more elaborate or larger 'head' it will be necessary to attach the mask to a simple 'hat'. The first stage in construction will be to make the 'hat'. Pupils should work in pairs to measure the circumference and distance across each other's heads. The hatband needs to sit at the top of the pupil's forehead. The hat frame is made first from strips of card that make a 'cap' on top of the pupil's head. When measuring remember to allow the necessary overlap for joining. The animal head or headdress design is then adapted to fit the pupil's measurements and incorporates the face mask component. The animal head is then attached to the hatband, taking care to ensure that the eye and nose holes are correctly located.
- The mask could be made from papier mâché that has been moulded onto a balloon blown up to a similar size as the pupil's head. This will give a realistic curve to the mask, which will fit the face better and allow some features to be built up in layers of papier mâché. However, remember that the papier mâché will take a long time to dry out, so prepare them well in advance.
- There are a large number of 'making' activities in which pupils may adapt the size of a standard design to fit themselves or some particular situation, and which require the pupils to measure accurately using standard or non-standard units. Textile or card bags can be made as props for a drama activity, or costumes created that are based on simple t-shirt and dress designs, with the pupils adapting the design to their own measurements. Glove puppets can be made to fit the pupils' hands or stuffed toy designs rescaled to suit a particular purpose. Preparing classroom displays can be used as an opportunity for pupils to apply ideas of ratio to scale up a picture or rough plan to match the size of the actual display board.

Activity 5

Making cakes and biscuits for the school fair

Key concepts
This activity could cover all aspects of the design cycle. Pupils can start with some market research to decide what type of cakes or biscuits are appropriate to sell at the school fair, potentially developing their own recipes and planning their

production system. Making the cakes will involve the use of appropriate tools, the consideration of alternative methods of mixing and measuring, enhancing the appearance of the finished product and evaluating the importance of following correct procedures for food safety and hygiene. The production process and the final product can be evaluated, including the amount of profit made for school funds. Within the activity pupils have the opportunity to learn about the ingredients used and if appropriate the materials used in any packaging.

The mathematics

Pupils will use aspects of data handling in the first part of the activity when they carry out market research to determine the most popular type(s) of cakes. Data from surveys conducted can be arranged in tables and graphs and interpreted to draw conclusions about the types and quantities of cakes to be produced. If the pupils use a standard recipe to make cakes or biscuits they will need to work with ratio and proportion to calculate the quantity of ingredients required. Tests must be conducted to see how many cakes are made by certain quantities of ingredients, particularly if they devise their own recipes. This will be an opportunity to work with the concepts of average and range. Calculations will be undertaken to scale up ingredients to the quantities required to make the number of cakes needed for the school fair. There will be extensive opportunities to practise measuring mass and capacity to an appropriate level of accuracy. The calculation of costs and setting sale prices for the cakes will involve solving 'real-life' problems with money. Spreadsheets could be used with the children to calculate overall costs and profits from the production of different numbers of cakes. If pupils devise a production process to make the cakes then the stages in this can be represented on a flow chart or timeline. The critical steps can then be identified and a suitable number of children involved in each stage of the production. If packaging is designed to make the products more attractive or to keep them clean and fresh then this could involve work on shape and space, for example visualising and drawing 2D and 3D shapes. Work on simple transformations will also be possible if packaging is printed using IT applications.

Note: The complete activity as described here would take up a significant length of time, so the steps marked with an asterisk could be treated as optional and answers given to the pupils.

*Step 1
Work together to examine the way a number of types of cake are made and what they taste like by producing them in small quantities (e.g. cup cakes, sponge cakes, butterfly cakes, cereal and chocolate cakes, which are decorated using icing, glacé cherries, hundreds and thousands, etc.).

*Step 2
A survey is conducted to identify the most popular type of cakes among children and parents.

D&T: Creating with numbers

*Step 3
The pupils design their own cakes and make them.

Step 4
The cakes are evaluated by a 'panel' of tasters and rated on a scale of 1 to 5 for various criteria (e.g. how they look, how they smell, how they taste, size, value for money, healthy eating). Enter the scores in a table, sum the scores and select the best one or two to be made for the fair.

Step 5
Help children estimate the number that might be sold based on data from previous years and by surveying a sample of children and parents. See if they can calculate the quantity of ingredients needed to make that number and how much this will cost. Let the group decide on the price of the cakes and calculate profits.

Step 6
Design the process for making the required number of cakes and try to split this into separate steps that can be completed by different groups of children. Represent the stages in a flow chart or timeline.

*Step 7
Decide what type of packaging is required and let the children design and make it. Alternatively you could purchase this ready made.

Step 8
Make the cakes together and sell them at the school fair. If children run the cake stall they will get extra opportunities to practise their calculation skills with money.

Step 9
Evaluate the project considering the success of the cake designs, the production system and the number of cakes sold on the day.

Extension activities
- You could make the activity into an 'enterprise' in which the pupils take on roles like manager or secretary for their group, and where they plan their own sales campaign. Posters to advertise the cakes before and during the fair can be made. Packaging can be designed and made to sell packs of five or six cakes. Remember to add advertising and packaging costs into the procedure for calculating the profit.
- Keeping full accounts of the production process will give extended practice at calculations involving money. This can encourage the pupils to look for ways of checking their own work and there will be instances when different calculations should result in the same answer.

- Ask pupils to organise a rota for children and adults running the cake stall using a timeline. See if they can devise ways to communicate this clearly to all those involved.
- Smaller-scale projects might involve providing refreshments at the school sports day or the Christmas play, where the pupils calculate how many cups of tea or glasses of orange squash will be needed and plan the purchase of the required materials.

Activity 6

A tasty, healthy lunch

Key concepts
This activity focuses on developing, planning and communicating ideas about the components of a healthy packed lunch. The problem of healthy lunches can be raised through discussion with pupils in which the conflict between the items that they would like to eat and those that make a healthy meal can be identified. By drawing on their own and other people's experiences they consider the sorts of foods that can be put into a healthy packed lunch.

Provide the pupils with information about what constitutes a healthy diet. For younger children you may quite simply identify actual food groups that should be included (e.g. fruit and vegetables; bread, cereal, rice and pasta; meat, poultry, fish, dry beans, eggs and nuts; sweets and dairy products). For older pupils the information might be given in terms of the quantities of major nutrient groups required (protein, carbohydrate and fats – possibly extending to consider fibre and energy). The problem could be made more wide-ranging by considering ways of packaging the lunch and designing advertising to promote it.

The mathematics
The identification of different foods and the nutrients they contain is part of a classification system that will be most clearly communicated and explained using a tree diagram. Finding and displaying information about a healthy diet and the dietary constituents of foods likely to be part of a packed lunch will involve the data-handling activities of collecting information and interpreting tables and charts. Converting information from one form to another, for example converting the amount of carbohydrate per 100 g of bread to the amount in one slice, will require some measuring (the weight of the slice of bread) and calculations with decimals. It could also include the use of a calculator. Comparing the recommended daily intakes of protein, carbohydrate and fat in a particularly healthy lunch with whatever else the pupil is assumed to eat during the day will involve further calculations. These could include the use of ratio and percentage.

D&T: Creating with numbers

Step 1
Pupils need to be aware of what is meant by healthy eating and the ways in which diet can affect our health. In the two weeks before the activity starts the pupils could be asked to bring in labels from a wide variety of food products. Make sure that similar information is available on the other ingredients likely to be suggested for the lunch like various fruits and bread.

Step 2
Pupils should be shown a classification tree that starts with the key nutrient groups – protein, carbohydrate and fat – and extends to some individual foods. For example, showing them the quantity of each nutrient group in 100 g of bread.

Step 3
Next give the pupils the daily recommended intakes of each nutrient group and through discussion establish what might be suitable minimum targets to eat at lunchtime.

Step 4
Pupils should then try to design a lunch that they would like to eat. Ask them to list in a table the contribution that each part of the lunch makes towards the target quantities of key nutrients. This will involve reading the food labels carefully and making calculations based on the quantity of each food included. These calculations will involve the addition, subtraction, multiplication and division of numbers with one or two decimal places. Some pupils will need to use a calculator to complete them.

Step 5
Finally the chosen lunch packs can be listed and used in a display where different lunches can be compared using graphs. These could show the actual quantities of protein, carbohydrate and fat in the lunch compared to the targets set, or the percentage of the total recommended daily intake.

Extension activities
- The pupils can identify the information on the food labels, display the information graphically and compare some typical products. They compare the food value of fruit and vegetables with the manufactured food products and write a report on their contribution to a healthy diet.
- The lunches designed could be 'road tested' and evaluated on a number of criteria. The resulting data should be analysed using graphs and charts and added to the display on healthy eating. The highest scoring lunches could be publicised in a school newsletter.

8 Art
Keeping things in proportion

Art has a number of defining interrelated characteristics including:

- *Personal elements* – perception of, and response to, the visual world, personal awareness and identity;
- *Visual elements* – ideas of beauty, the awareness of the range of representations that can be used to express responses to the visual world and personal feelings;
- *Technical elements* – acquisition of necessary and useful skills and techniques, knowledge of tools, materials and procedures.

These elements can be united to make works of art in a variety of media, which can express a multitude of thoughts and feelings through a variety of symbols. The artist reflects on experience, and communicates the ideas and emotions it engenders to other human beings.

Where is the mathematics in art?

Because art is a particularly visual, intuitive form of knowing that uses non-discursive ways of communication, it does so through symbols that are independent of, and often inexpressible through, language. There appears to be four interdependent areas in art where mathematics can usefully contribute to the visual representation of thoughts and feelings and the way in which they are organised. Firstly, mathematical concepts such as shape and proportion are seen in the natural and human environment and recorded. Secondly, mathematical ideas can be used in the production of creative work: mathematical shapes and pattern are used as motifs in the creative act itself and geometric figures, proportion, symmetry and asymmetry, measurement and line all play important parts in self-expression. Thirdly, to gather materials and review progress effectively, artists need to quantify through estimation the amounts of materials and space needed and to measure and confirm these estimates. Ordering skills will be needed to bring sets of materials together correctly and to undertake techniques in consecutive stages. Finally, there will be the qualification of other's artwork and the need to discover what mathematics was used, why it was used and whether or not it has been successful.

Art: Keeping things in proportion

Activity 1

Weaving

Key concepts
Weaving offers many opportunities for studying surface pattern in the warp and weft and in vertical and horizontal lines of weave. In studying the design and making of patterns pupils can examine various types of weave and identify their construction as a prelude to making their own patterns at a variety of scales.

The mathematics
The activity develops the mathematical concepts of horizontal (i.e. that which is parallel to the plane of the horizon and at right angles to the vertical) and the vertical, which is perpendicular to the horizontal.

Step 1
Children choose a variety of woven cloth.

Step 2
They identify the warp and weft, tracing it with their fingers.

Step 3
They use the language of 'horizontal' and 'vertical' to describe the elements of the cloth that they are tracing.

Step 4
They explore the effect on colour and texture of using different threads horizontally and vertically.

Step 5
They record their findings as annotated sketches in their sketchbooks.

Step 6
Pupils can then use these sketches to produce their own woven fabrics.

Extension activities
- Let the children explore the use of pattern in various examples of woven art, analysing the arrangement of threads and colours and how colour and line have been used.
- Investigate woven spiral shapes with the children and examine how these relate to the centre of the work.
- Look at examples of circular weaving and explore the relationship of perimeter to radius.

Numeracy and Mathematics Across the Primary Curriculum

Activity 2

Comparing surface area with mass

Key concepts
Children enjoy working with materials such as clay on account of their malleability. A particular weight of clay can be metamorphosed into a great variety of shapes.

The mathematics
Using a fixed amount of clay to produce pots involves measuring and calculating surface area and capacities. A ball can be rolled out very thinly and shaped into a cube or cuboid, or a stubby thumb pot can be stretched into a very thin, fragile form. By experimenting in this way children will be exploring which shape is the easiest to make and has the greatest capacity.

Step 1
Set the challenge to see how many pot shapes can be made using the same mass of clay.

Step 2
Cut a number of blocks of clay and weigh them, paring or adding clay until they are the same mass.

Step 3
Make a different pot shape with each unit of clay: tall, shallow, round, square, triangular, etc.

Step 4
Measure the amount of liquid each will hold by filling them to the brim with water.

Step 5
Cut each pot into parts that can be flattened to measure surface area.

Step 6
Ask the children to consider which is the most efficient shape for holding water. Why do we not have more square pots?

Extension activities
- Consolidate the concepts of capacity and surface area of solid forms by asking pupils to design pots that are cylindrical, cuboid or irregularly shaped.

Art: Keeping things in proportion

Activity 3

Developing design through number

Key concepts
Many patterns in a variety of cultures are based on manipulating numbers to produce pattern and symmetry. This is particularly so in Islamic art where the use of representational motifs such as human beings or animals was once prohibited.

The mathematics
The activity is based on producing digital roots (i.e. the simplest form of a number after continuous adding) from multiples of numbers – in this case times tables: Multiply any two numbers together (e.g. $4 \times 3 = 12$). Add the digits of their product together to find the digital root: $1 + 2 = 3'$. As well as making pleasing and complex patterns, this activity allows pupils to see patterns in number and to predict the continuing effects of the numbers on the pattern being formed.

Step 1
Tell the children that they are going to find out how they can use digital roots to make a pattern.

Step 2
Ask the pupils to choose a times table, say 6, and write down the resulting products. They then add the digits of each of those products until they achieve the digital root.

Number bond	Digital root
6	6
12	3
18	9
24	6
30	3
36	9
42	6
48	12 = 3
54	9
60	6
66	12 = 3
72	9

Step 3
Pupils identify the repeated patterns being formed and predict what the next three in the sequence will be.

> *Step 4*
> Using 1 cm square graph paper, they plot the sequences of digital roots along the line turning right angles after each individual root and always turning in the same direction. So, start by drawing a vertical line upwards from a given point (6 cm), then move through a right angle to the right (3 cm), then through a right angle to the right (9 cm) and so on until a pattern is formed. They continue until the pattern is complete, i.e. it forms an unbroken line.
>
> Plotting digital root sequencs

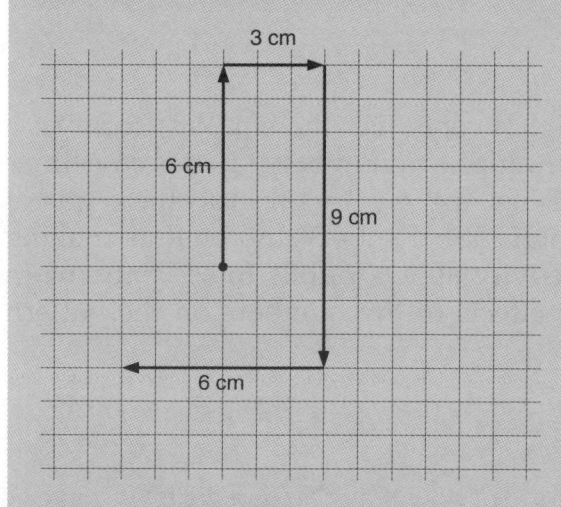

Extension activities
- Help children to explore patterns achieved using other multiplication tables.
- Together research patterns made by square or cubed numbers.
- Use rangoli patterns and explore other design motifs from Celtic art and Roman mosaics as a stimulus for creating shaped tie-dye designs.
- See what can be produced using 'repeat', 'blow up' and other facilities in computer art.
- Look at existing printed forms and see if children can identify repetition.
- Challenge children to identify examples of rotation, translation, reflection or enlargement in the pattern.

Activity 4

Reflective symmetry and positive or negative shapes

Key concepts
Reflective symmetry is an important effect in many forms of design. This activity uses the concept at its simplest – that of a mirror image. This can be the basis for a number of design activities and can be extended to incorporate a variety of materials.

Art: Keeping things in proportion

The mathematics
Reflective symmetry, also known as line, bilateral or mirror symmetry, is essentially about the correspondence of points that are on opposite sides and equal distance from a line. In this case the line appears to divide a shape into equal halves.

Step 1
Tell the children that they will be exploring how to make positive and negative symmetrical shapes.

Step 2
Using black and white paper ask pupils to cut pieces measuring approximately 20 cm by 5 cm. (Check that scissors are being used safely throughout this activity.)

Step 3
They then take one of the pieces of white paper and cut a series of V-shaped notches (wide end of the V to the edge of the paper) down one of the 20 cm sides. Each notch should be 1 cm wide and cut 3 cm into the width of the paper. Make sure they leave 1 cm space between the notches.

Step 4
Ask the pupils to put the piece of paper they have cut into onto a piece of black paper/card and reassemble all the pieces into the original shape.

Step 5
They should then take each cut-out notch and rotate it 180 degrees so that they lie reversed along the cut-out edge of the paper. Ask the children to paste them down.

Extension activities
- Start with different shaped pieces of paper and cut out more complex shapes, or try a variety of colour combinations.
- Children can make a wax crayon mirror image by folding a piece of paper in half and putting a thick layer of chalk on one half. Over this a thick layer of crayon is laid down. Fold the paper in half with the crayon on the inside and at the bottom. Draw a motif or shape on the plain side that faces upwards using a biro or a pencil. Open the paper up and the shape will appear as a reflection on both sides of the folded paper.

101

Activity 5

Proportion

Key concepts
While developing understanding of the work of artists, crafts people and designers, and applying this knowledge to their own work, pupils can explore the idea of proportion in paintings: the relation of one part of the piece to other parts, or to the whole of the painting.

The mathematics
Proportion in mathematics is closely related to the concepts of fractions and ratio. When corresponding pairs are in the same ratio they are said to be 'in proportion'. This activity explores a number of elements of a piece of artwork and how they relate to parts of the painting or to the piece as a whole.

Step 1
Pupils informally examine a painting for key features such as people, buildings or trees.

Step 2
They measure the distance between these elements to see if any pattern of lengths emerge (e.g. halves, thirds or quarters).

Step 3
The children can then overlay the painting with a grid of squares, or horizontal or vertical lines to see if anything becomes apparent.

Step 4
Pupils take the height of the key figure in the painting and use it to measure the ratio with any other feature.

Step 5
See if they can find a 'unit' – a common measurement of distance that seems to be embedded in the features.

Extension activities
- Ask the children to explore the proportions of a standing pupil or the human face.
- Together examine the proportions of the human figure in paintings by various artists.
- Use enlargements – transformations that map a shape onto a similar shape from a centre using a scale factor – exploring whether the proportions change.

Art: Keeping things in proportion

Activity 6

The 'golden rectangle'

Key concepts
Composition is the arrangement of shapes in a painting. Often artists use flat shapes such as squares, triangles or circles to arrange the features in their paintings. One shape often used to structure paintings is the 'golden rectangle' thought to be one of the most visually satisfying of all geometric forms. This activity explores the use of this shape in paintings.

The mathematics
The golden rectangle is formed by the ratio of 1.618 and is the ratio between Fibonacci numbers. This sequence of numbers 1, 1, 2, 3, 5, 8, 13, 21, 24, 34, 55 is arrived at by adding the previous two numbers. The sequence is named after the Italian mathematician, Fibonacci, also known as Leonardo of Pisa (1170–1230). The ratio between the numbers is 1.618. The activity asks the pupils to make a golden rectangle and to use it to identify whether any aspects of the painting fit into it.

Step 1
Pupils draw a rectangle of 1:1.6 ratio, say 3.0 cm by 6.4 cm, on card and cut out the inside of the shape to make a 'golden rectangle finder'. The dimensions of the frame around it are a matter of taste and convenience.

Step 2
They then work out the measurements using a matrix. Once they have spotted the pattern and recognised the relationship between the numbers, the matrix should be simple for them to complete.

Side 1 (cm)	Side 2 (cm)
1.0	1.6
2.0	3.2
3.0	6.4
6.0	
10.0	

Using a matrix to calculate measurements

Step 3
Several finders of different dimensions can be made.

Step 4
Pupils use the golden rectangle finder to record as many examples of it being used as they can. The painting is pinned to a wall and they hold the golden rec-

tangle finder at arm's length either vertically or horizontally to see how many of them be can detected within the painting. The findings are recorded in a matrix:

Feature	Golden rectangle used?
Group of people	Yes
Whole shape of painting	Yes
House shape	No

Matrix of findings

Extension activities
- Help the children to explore uses of the golden rectangle in other visual forms such as sculpture and pottery.
- See what shapes make up different compositions (e.g. circles, triangles or squares). Use a variety of sizes of these shapes on OHP acetates.
- Together study artists' use of cubes, spheres or cones in the composition of paintings.

Activity 7

How can we measure colour?

Key concepts
Colour is a fascinating area for children to think about as it comes in so many shades. Children are used to the *ad hoc* mixing of colour and this activity endeavours to add a structure to this process.

The mathematics
Trying to quantify colour introduces aspects of informal measurement, using improvised tools and/or invented non-standard units. In this activity the units are 'paintbrush-fulls'.

Step 1
Choose a colour (e.g. green) and ask the children to collect samples of it from natural and human environments (e.g. grass from the playing field, or a green hair ribbon). Ask them to sort their samples from light to dark.

Art: Keeping things in proportion

> *Step 2*
> They experiment in matching the colour with paint by creating different shades. They should lighten or darken the original colour from their paintboxes by adding touches of any other colour.
>
> *Step 3*
> Get them to record their 'recipes' (e.g. green + one brush of yellow) and cover a square of paper with the resulting colour. Ask the group to then put these in order from lightest to darkest.
>
> *Step 4*
> The children then use this as an informal measuring scale to say how green something is.

Extension activities
- Help the pupils to make abstract patterns using 'colour' families.
- Use the colour families to make a patchwork quilt painting. The colours can be repeated and this can be recorded algebraically.
- Use the primary colours – yellow, red and blue – to make new colours. Decide on the 'unit' amount of colour (one paintbrush-full) and record this using number codes as in Figure 8.1. What other combinations are possible using three colours and three units of colour?

Yellow = 1	1 + 1 + 2	1 + 1 + 3	1 + 2 + 3
Red = 2	2 + 1	2 + 3	2 + 2 + 3
Blue = 3	3 + 1	3 + 3 + 1	3 + 3 + 2

Figure 8.1 Mixing colours and units of colour

References

Askew, M. (1999) 'It ain't (just) what you do: effective teachers of numeracy', in Thompson, I. (ed.), *Issues in Teaching Numeracy in Primary Schools*. Buckingham: Open University Press.

Buxton, L. (1981) *Do You Panic About Mathematics?* Oxford: Heinemann.

Copeland, T. (1992) *Mathematics and the Historic Environment*. London: English Heritage.

Crook, J. and Briggs, M. (1988) 'Bags and baggage', in Pimm, D. and Love, E. (eds), *Teaching and Learning School Mathematics*. London: Hodder and Stoughton.

Department of Education and Science (DES) (1982) *Mathematics Counts* (Cockcroft Report). London: HMSO.

Department for Education and Employment (DfEE) (1999) *The National Numeracy Strategy: Framework for Teaching Mathematics from Reception to Year 6*. London: DfEE.

Department for Education and Employment (DfEE)/Qualifications and Curriculum Authority (QCA) (1999) *The National Curriculum: Handbook for Primary Teachers in England: Key Stages 1 and 2*. London. HMSO/QCA.

Haylock, D. and Cockburn, A. (1997) *Understanding Mathematics in the Lower Primary Years*. London: Paul Chapman.

Hiebert, J. and Carpenter, T. P. (1992) 'Learning and teaching with understanding', in Grouws, D. A. (ed.), *Handbook of Research on Mathematics Teaching and Learning*. New York: Macmillan.

Her Majesty's Inspectorate (HMI) (1985) *Mathematics 5 to 16*. London: HMSO.

Hughes, M., Desforges, C. and Mitchell, C. (2000) *Numeracy and Beyond: Applying Mathematics in the Primary School*. Buckingham: Open University Press.

Mallett, M. (2002) *The Primary English Encyclopaedia: The Heart of the Curriculum*. London: David Fulton.

Schools Curriculum and Assessment Authority (SCAA) (1997) *The Teaching and Assessment of Number at Key Stages 1–3*. York: SCAA.

Index

activities 14
art 96–105
   mathematics in 96

bar charts 7, 20, 21, 23, 33, 39, 44, 49, 62, 63

Carroll diagrams 17, 18
charts 33, 44, 46, 50, 94
chronological sequence 6, 25, 27
chronology 5, 25, 26
climate 37
cognitive structures 8, 57
colour 104–5
connected learning 3

dance 60–2
data handling 20, 23, 58, 62, 94
design and technology 8, 80–95
   mathematics in 82
diet 20–1, 94–5
directional language 37

Earth 19
ELPS framework 3
enactive experiences 3
English language 43–55
   mathematics in 44
environment 40–2
extension activities 14

fiction 43–6
flow chart 26, 38, 46, 89

Gantt charts 44, 45, 46, 50, 88
generations 29
geography 12, 35–42
   mathematics in 35
geometry 31, 56–68, 82
'golden rectangle' 103–4
graphs 16, 20, 21, 24, 39, 44, 50, 92
gymnastics 67–8

heat 23
history 5, 6, 25–34
   mathematics in 25
   number lines in 6

information processing 9

key concepts 14

learning, transfer of 4
learning objectives 10
line charts 23, 44
line graphs 22

mapping 35
map reading 66
mass 98
materials, technology 17–19
mathematical
   activities 13
   concepts 7, 14, 35
   confidence 10
   learning 10
   modelling 1–2, 4, 20
   patterns 73, 75

Numeracy and Mathematics Across the Primary Curriculum

mathematical – *continued*
   understanding 6–7, 10–12
mathematics, attitude to 12
motivation 12–13
music 69–79
   mathematics in 69

National Curriculum 9, 14, 16, 81
National Numeracy Framework 9
negative numbers 6
non-fiction 43, 46–51
number line 6, 28
number order 6
numeracy 8

orienteering 64–7

patterns 7, 99–100
physical education 8, 56–68
   mathematics in 57
pie charts 20
poetry 44–5, 51–5
polar coordinates 37–8
problem solving 1–2, 8, 9, 11, 14, 66
problem-solving cycle 5
problems, realistic 2
proportions 102

ripple diagram 41
rhyme 54–5
rhythm 54, 61, 69–71, 75

Schools Curriculum and Assessment Authority 2, 10
science 15–24
   mathematics in 16
self assessment 13
situated cognition 4, 57
skills 7, 9, 10, 52, 57, 78, 80, 81, 83, 96
   creative thinking 9
   cycle 11
   enquiry 4–5, 9, 16
   evaluation 9
   getting 11
   reasoning 9
   thinking 9
   using 11
strength 85–7
surface area 98
symmetry 31–2, 45, 57, 67, 70, 79, 90, 96, 100–1
   buildings 31
time line 6, 27, 28, 29, 34, 75, 89
Titanic 46–8
transfer of learning 4
transmission model 8

Venn diagram 17, 18, 32, 33, 44, 49

weather 38
weaving 97